Jacob Ennis

The Origin of the Stars and the Causes of Their Motions and Their Light

Jacob Ennis

The Origin of the Stars and the Causes of Their Motions and Their Light

ISBN/EAN: 9783337025366

Printed in Europe, USA, Canada, Australia, Japan

Cover: Foto ©berggeist007 / pixelio.de

More available books at **www.hansebooks.com**

THE

ORIGIN OF THE STARS,

AND

THE CAUSES OF THEIR MOTIONS AND THEIR LIGHT.

JACOB ENNIS,

PRINCIPAL OF THE SCIENTIFIC AND CLASSICAL INSTITUTE, PHILADELPHIA;
FORMERLY PROFESSOR OF THE NATURAL SCIENCES IN THE
NATIONAL MILITARY COLLEGE.

NEW YORK:
D. APPLETON AND COMPANY,
443 & 445 BROADWAY.
1867.

Entered, according to Act of Congress, in the year 1866, by
JACOB ENNIS,
In the Clerk's Office of the District Court of the United States for the Eastern District of Pennsylvania.

PREFACE.

No inquiry is more congenial to the human mind than the origin of man and of the great creation around us. Hence the hero of Milton's grand poem relates how, on first awakening into being, he began these queries—

> "Thou sun, said I, fair light,
> And thou enlightened earth so fresh and gay,
> Ye hills and dales, ye rivers, woods, and plains,
> And ye that live and move, fair creatures, tell—
> Tell if ye saw, how came I thus, how here."

In pursuing these investigations in a scientific point of view, the greatest advance made by the ancients was the idea that our earth is round like a globe. This was an indispensable step toward the discovery of its planetary or stellar nature. In modern times there have been many theories formed to account for the different phenomena of creation, some explaining one part of the process, and some another. Ten of these are slightly sketched in the forty-second section, but the entire volume is employed in advocating four of those theories, each one occupying one of the four parts of the work.

The First Part begins, like the Book of Genesis, with the creation of light. There seems always to have been a vague opinion that the light of the sun is produced by a great burning. Newton and Laplace believed this; and Davy believed that our planet, the earth, was formerly in a fused, glowing condition by the chemical union of its simple elements. But for the last half century it has been the very decided opinion of scientific men, that chemical action or burning cannot account for the heat of the sun, because that body does not contain a sufficient amount of fuel. Even Newton and Laplace and the older philosophers never entered into any extended arguments on this subject. They simply expressed their opinion, fortified with one or two reasons. This volume is the first serious attempt to take up the old chemical theory of stellar light and heat, and to give it a full, elaborate discussion. I have tried to do this in a way to interest and benefit not simply scientific men, but the general mass of intelligent readers. Even if the latter fail to come to my decision, they will nevertheless become familiar with a large number of facts in science. For it is on the collection, the description, and the explanation of facts, that the argument depends.

The Second Part contains a new theory of the force which has prolonged the light and heat of the sun through the vast duration revealed by geology. This theory is introduced by the facts on which it is founded, and these are stated in the form of answers to the single objection against the chemical theory of stellar light. The present extravagant estimate of that objection led me to take this course.

The Third Part, I think I may fairly say, demonstrates that the origin of the stars was from a condensation of matter previously in a very rare or gaseous condition. This idea is not new; but the demonstration of its truth, long desired, has never before been given. This part also now first explains the origin of a large number of astronomical phenomena, such as the reason why the four exterior planets have all the satellites except one; and why, of the four interior planets, the earth alone has a satellite.

The Fourth Part completes the argument already begun in the third, to show that Gravity is the force which originally gave their motions to the stars.

Thus we now know, without any doubt, the origin of the great globes which roll through space, and we know the force which sent them onward with such astonishing velocities as a thousand miles a minute; the velocity of our earth, for instance, being greater than that.

We also know what causes the light and heat of the

sun and of the other stars; and we know why many stars, like our earth, have lost their light, and shine no more. These truths about their light and heat may not at first be believed by some, but I have the strongest confidence that before long they will not be doubted.

PHILADELPHIA, *April* 6, 1866.

CONTENTS.

                                                                            PAGE

PREFACE.. 3

PART I.

THE CAUSE OF THE LIGHT AND HEAT OF THE SUN AND THE OTHER FIXED STARS.

SECTION

I. INTRODUCTION.. 11
II. More than forty points of resemblance between our Earth and the other planets........................... 12
III. More than twenty facts proving that the fixed stars are suns. 17
IV. Suns, planets, and moons are bodies of the same nature... 23
V. Evidences from geology and kindred sciences that the Earth was once self-luminous like the Sun................ 28
VI. Evidences from chemistry that the Earth was formerly self-luminous... 57
VII. The Sun... 73
VIII. The fixed stars..................................... 102
IX. Review of evidences................................. 164

PART II.

THE FORCE WHICH SO GREATLY PROLONGS THE LIGHT AND HEAT OF THE SUN AND STARS.

X. The single objection to the chemical theory of stellar light and heat is founded on three groundless Assumptions... 170
XI. The First Assumption................................ 172

SECTION	PAGE
XII. The Second Assumption	176
XIII. The Third Assumption	182
XIV. The facts explained by the chemical theory of stellar light and heat	206
XV. The theory of the force which prolongs the chemical action of the stars	214

PART III.

THE ORIGIN OF THE STARS.

XVI. The former universal diffusion of all matter	216
XVII. The necessity of rotation in a contracting nebulous mass.	220
XVIII. The velocity of rotation	226
XIX. Retardation from friction	231
XX. From a rotating nebulous globe gravity separates a cluster of planets and satellites	239
XXI. The satellites	243
XXII. Why the equatorial velocity of the Sun is less than the orbital velocities of the planets; and why the equatorial velocities of the planets are less than the orbital velocities of the satellites	251
XXIII. Why the Sun has been retarded in its rotation so much more than the planets	255
XXIV. The distances of the planets from the Sun, and of the satellites from the planets	256
XXV. Why the four exterior planets have all the satellites but one: and why of the four interior planets the Earth alone has a satellite	261
XXVI. Why the Sun's nearest attendant is so far off; why Saturn's nearest attendant is so near; why there is no planet between Mercury and the Sun	263
XXVII. The peculiarities of the Saturnian system	264

| SECTION | PAGE |

XXVIII. Why the planets are so much more distant from one another than the satellites..................... 267
XXIX. The peculiarities of the Mundane system........... 268
XXX. Why the planets and satellites rotate from west to east.. 274
XXXI. The inclination of axes of rotation; the inclination of the orbital planes of secondaries to the equatorial planes of their primaries; the retrograde motions of the satellites of Uranus..................... 278
XXXII. The eccentricities of the planetary orbits........... 287
XXXIII. The asteroids, meteorites, and comets.............. 292
XXXIV. The double, triple, and multiple stars............... 295
XXXV. The various forms of nebulæ or sidereal systems...... 299
XXXVI. A central Sun?................................. 306
XXXVII. The proper motions of the fixed stars.............. 310

PART IV.

THE FORCE WHICH GAVE MOTION TO THE STARS.

XXXVIII. Gravity originally gave their motions to the stars..... 316
XXXIX. A summary of the facts in the solar system whose origin is accounted for by gravity.............. 321
XL. A comparison between the evidences that gravity now holds the planets in their orbits, and the evidences that gravity originally gave them their motions in their orbits.................................. 326
XLI. The opinions of scientific men hitherto on the cause of the motion of the stars......................... 331
XLII. Modern theories of creation....................... 343
XLIII. The natural history of creation.................... 359
XLIV. The creation and government of the world by special forces.—Divine Providence 365

APPENDICES.

	PAGE
Appendix I. A mathematical process....................	367
" II. An answer to a mathematical objection.............	368
" III. Velocities necessary in the fixed stars to prevent their coming into collision........................	375
" IV. Catalogue of writers on the cause of stellar motion as given in the forty-first section.............	376
" V. Laplace's writings on the nebular theory..........	378
" VI. The depressions caused by the spots on the surface of the Sun..................................	383
Note to Section XVII., on the necessity of rotation.............	383
" " XXIX.................................	384
Personal Index...	387
Index of Stars...	389
General Index..	391

PART I.

THE CAUSE OF THE LIGHT AND HEAT OF THE SUN AND THE OTHER FIXED STARS.

SECTION I.

INTRODUCTION.

THE object of this small volume is to show the origin of the stars, and the causes of their motions and their light. The reader may at once inquire with surprise, How is it possible for us to learn the origin of the distant stars? How can we ascertain the cause of their shining; and how can we learn what mighty force gave them their rapid motions? One of our chief sources of information, though not by far our only source, arises from the fact that our earth is a star, and moves among the stars, and that once it shone with all their splendor. Here, then, we live and move on a star, and we have an opportunity to inquire into its past history and its ancient origin. Here we can examine its constitution, and find out how its elements could blaze like the sun. Therefore in our great undertaking we must lay the foundation by proving that our earth is a star; not only a planet appearing beautifully like Venus and Jupiter in the evening sky, but really a fixed star that once shone by its own independent light. The more extensively we see the resemblance between our earth and the other stars, the greater will be our confidence in applying our terrestrial knowledge to the celestial bodies; and hence we must be very familiar not only with a few, but with all those facts which are common to our earth and to them. Our plan will be first to enumerate all the coincidences between our earth and the other planets; and then to show that between planets, suns, and fixed stars there is no essential distinction, that all belong to the same general class of bodies, and that their history and their origin have been the same. This must be the corner-stone of our edifice.

SECTION II.

MORE THAN FORTY POINTS OF RESEMBLANCE BETWEEN OUR EARTH AND THE OTHER PLANETS.

1. They are all very great bodies, their diameters ranging from 2,950 to 88,600 miles; that of our earth is 7,912 miles, and therefore of an intermediate size.

2. They are all round bodies; this is evident by simple observation.

3. They are all slightly oblate, being flattened more or less at the poles, so that their polar is less than their equatorial diameters.

4. They are all opaque bodies, obstructing the light of the sun, and therefore the shadows of them all eclipse their satellites or moons.

5. They all reflect light. The mildness of their reflected light causes the planets and the moon to look so beautifully by night. The reflection of the light from our earth is seen on the dark portion of the moon. From the moon it is reflected back a second time to ourselves.

6. The planets are like our earth in having matter in the solid form.

7. They also have matter in the liquid form.

8. And they have matter in the gaseous or vaporous form.

9. They have inequalities on their surfaces, the same as our earth has hills and mountains.

10. They are surrounded with atmospheres.

11. They have clouds in their atmospheres, and these clouds are seen like ours to change their forms.

12. At least one of them, Mars, has its polar regions, when turned away from the sun, covered with a white appearance like those of our earth, which are whitened by snow in winter. If there were no snow on the polar regions during the many weeks of wintry nights, those regions should appear especially dark. Snow and clouds presuppose the existence of a liquid similar to water; an atmosphere implies matter in a gaseous form; and mountainous inequalities of surface tell of matter in a solid form.

12. The atomic force of cohesion acts there the same as here, otherwise there could be no solids.

14. Also the atomic force of repulsion reigns among the planets, for without this there could be no gaseous atmospheres.

15. The force of gravitation manifests its activity among the planets by moulding them in a round form, and by retaining their satellites in their orbits.

16. The centrifugal force shows its power among the planets precisely as here, by giving them their oblate forms, and by counteracting gravitation.

17. The matter of the planets, like that of our earth, is endowed with the property of inertia; this is evident, because inertia is the foundation of the centrifugal force.

18. The planets with our earth are distributed one beyond another in distances from the sun in such a manner as to prove a family likeness and a common origin. Each

one, with the exception of the first, is so placed that the interplanetary spaces are wider and wider the farther they are from the sun; and with the exception of the first and the last, each succeeding space is about double the preceding. The first and the last are exceptions, as we shall see, on account of the peculiar beginning and ending of the cause which assigned the planets their positions.

19. The planets, like our earth, revolve around the sun.

20. The motions of all are exceedingly rapid, that of the slowest being more than 12,500 miles an hour.

21. Their relative velocities show a common origin; the farthest distant from the sun being always the slowest.

22. Hence the more distant a planet is from the sun, the longer is its year or PERIOD of revolution.

23. There is a perfect mathematical relationship between their periods and their distances from the sun. Hence by knowing the distance of a planet from the sun, we can tell its period; or knowing its period, we can tell its distance. This relationship is expressed in these terms: The squares of the periods of any two planets, our earth included, are to each other in the same proportion as the cubes of their distances from the sun.

24. The orbits or paths of the planets around the sun are nearly in the same plane.

25. They all move around the sun in the same direction, from west to east; the same as the rotation of the sun on its axis.

26. Their orbits are all ellipses, differing only slightly from circles.

27. These orbits are so disposed that the sun is always in one of the foci of the ellipses.

28. Hence all the planets are sometimes near and sometimes farther from the sun. By their distance ordinarily is meant their mean distance.

29. An imaginary line, called the radius vector, drawn from the planet to the sun, sweeps over equal areas in equal times.

30. The planets are all subject to perturbations; that is, to be slightly impelled in various ways out of their paths by each other's gravity.

31. The major axes of all the planetary orbits constantly though slowly change their directions, each one making a complete revolution after a very long period.

32. The minor axes of all the planetary orbits constantly though slowly change their lengths, each one vibrating to and fro within certain limits in a very long period.

33. The planes of all the planetary orbits vibrate to and fro in very long periods.

34. The planets, like our earth, rotate on their axes, giving to them all the changes of day and night.

35. These rotations are all from west to east.

36. Their axes of rotation are all more or less inclined to the planes of their orbits, thus imparting to them all the vicissitudes of the seasons.

37. All the planets larger than our earth have one or more moons.

38. These moons all rotate on their axes.

39. They all revolve around their primaries.

40. Their rotations are performed, as far as is yet known, once in one revolution.

41. Both their rotations and their revolutions are from west to east.

42. The revolutions of all the moons, like those of all the planets, are in elliptical orbits.

43. The planes of the lunar orbits are nearly in the equatorial planes of their primaries.

44. The lunar orbits, like those of their primaries, are subject to various changes. If the parallel were drawn

between the changes in the orbit of our moon and in those of the satellites, and also between their velocities, distances, periods, radii vectores, eccentricities, major and minor axes, then the coincidences between our earth and the other planets would be very greatly increased beyond our present enumeration.

45. Like our earth the other planets are more dense in their interiors than at their surfaces.

46. The planets are more dense the nearer they are to the sun, forming a regular progression of increasing density from Neptune to Mercury. The causes of the slight exceptions of Saturn and Venus will be explained on another page.

47. As our moon is less dense than the earth, so the satellites of Jupiter are less dense than Jupiter. The densities of the other satellites have not been ascertained.

In reviewing these points of resemblance between our earth and the other planets, it is wonderful to see not only how numerous they are, but how truly essential. They are not slight and small, but great and fundamental. They are such important things as their sizes, shapes, and densities; their power of obstructing and reflecting the rays of light; their possession of solids, liquids, and gases; their having atmospheres, clouds, and mountainous inequalities; their subjection to the forces of gravity, atomic attraction, atomic repulsion, inertia, and to all the laws of motion; their distribution at regular distances from the sun; their motions in ellipses, with all the various changes of these ellipses; their rotations on their axes; their revolutions around the sun in the same direction and nearly in the same plane; and their being accompanied by round moons all moving alike. If any thing in the wide compass of our knowledge is evident, then it is clear that our earth is a star like the planets, and the planets are worlds like our own. Essentially and fundamentally they are the same, though in minor matters each one has its own peculiarities.

SECTION III.

MORE THAN TWENTY FACTS PROVING THAT THE FIXED STARS ARE SUNS.

We now go on to prove that our earth is not only a star like the other planets, but a star like the other fixed stars. We shall prove, first, that the fixed stars are bodies like the sun; then that there is no natural distinction between suns and planets; and lastly, that there is no natural distinction between planets and satellites. The ordinary distinction between suns, planets, and satellites, is altogether arbitrary, and used merely for convenience in a system like our own to designate readily which bodies we mean. But we will soon prove that these several bodies graduate from one to another by insensible degrees; that when viewed largely we can make no clearly defined separations between them; and that therefore they all belong to the same general class. They are the same, in fact, as the other fixed stars.

The facts proving the fixed stars to be suns are the following:

1. They shine by their own independent light. This is evident from their vast distances. No reflected light from our sun could give them their present brightness.

2. The amount of their light is equal to that of our sun, and often much greater. This becomes evident from knowing their distances, and the quantity of their light that reaches our vision.

2. Their masses, or the amount of matter in some of them, has been ascertained, and they are similar to that of our sun, " neither vastly greater nor vastly less."

4. Like our sun they have dark spots on their sides; and

5. Like our sun they rotate on their axes. These two facts are satisfactorily made out by the periodic increase and decrease of the light of many stars. Such increase and decrease may be accounted for on the supposition that these stars have large dark spots on their sides, larger than those on the sun, and that they diminish the stellar light when turned in our direction. Different estimates have been made of the period of the rotation of our sun, varying from twenty-five to thirty days; and this difference in the estimates arises in great part from the movements of the spots. If they were fixed, the period of rotation could be determined precisely. The period of rotation of some stars is shorter, and of others longer, than that of our sun, and it is an important consideration that their periods are also uncertain, and this uncertainty may be accounted for, as in the case of the sun, by the movements of the spots. A very few are precisely regular in their periods, and hence we may infer that in these cases their spots are not like floating islands, but that they have become fixed and continental.

6. The dark fixed lines exist in the prismatic spectra of the stars as in that of the sun. This is a highly interesting fact, because it carries along with it a large number of other most instructive facts. It must, therefore, be explained. When the light of the sun is admitted into a dark room through a small opening, and the whole of this light is decomposed by a prism into what are called the seven prismatic colors, or the prismatic spectrum, then many fine dark lines are seen to cross this spectrum, if viewed by a suitable instrument. These are called fixed lines, or Fraunhoffer lines, after their discoverer. The cause of these lines was discovered by Kirkhoff in the following manner: If the prismatic spectrum of an ordinary lamp be

taken, then various fixed lines are also seen, and the number and position of these depend on the substance burning or vaporized in the flame. But instead of being dark they are variously colored, and these colors also depend on the substances in the flame. They can all be made dark, however, just like those in the solar and stellar spectra. For if another and far brighter lamp be made to shine through the former and through the prism, then the fixed lines of whatever color or position all become dark. Therefore the interpretation of the dark lines in the solar and stellar spectra is as follows: In the flames surrounding the sun and stars there are various substances similar to those in our earth. Here in our earth there are more than sixty different substances called chemical elements, also giving their own fixed lines in the artificial spectrum. In the sun there must be a much larger number of chemical elements, because the fixed lines of its spectrum are so exceedingly numerous. In the stellar spectra they are few, on account of the very small amount of their light. But why in the sun and the stars should all the lines be dark? Because behind the flames containing the vapors of the chemical elements there must be a much brighter light shining through them. This brighter light must come either from a solid or a liquid. From the great intensity of the heat, it cannot be a solid. It must therefore be a liquid—the liquid body of the sun and of the fixed stars. This liquid condition of the sun will allow currents as in our ocean, and these currents explain the movements of the spots, which spots, from their cracking and breaking and the evidences soon to be given, are certainly solids, but only temporary solids from the intensity of the heat.

These dark fixed lines in the solar and stellar spectra prove not only that the sun and fixed stars are bodies of a similar nature, but that they are bodies similar to our earth, whose elements are capable likewise of producing fixed lines in the spectrum. We are thus enabled to enumerate other points of resemblance between the sun and the fixed stars.

7. Both the sun and the fixed stars are composed of elements similar to our earth.

8. There is the most intense heat in the stars as well as in the sun.

9. There are envelopes of flames surrounding both the sun and the fixed stars. This agrees with Arago's previous and independent proof from the want of polarization in the light of the sun.

10. The bodies of the sun and of the fixed stars are liquid. This harmonizes with the conclusions of geologists that our earth below a comparatively thin crust is now in a liquid condition, and that formerly it was all liquid except here and there some floating solid spots, which at first were temporary and afterwards increased in size and number, and became the present permanent crust.

11. Matter in the sun and in the fixed stars can take the form of vapor. This is proved by the fixed lines, because they can be formed only by vapor.

12. Matter in the sun and the fixed stars can exist in the liquid condition. This is evident from the liquid nature of their bodies.

13. Matter in the sun and the fixed stars can exist in the solid form. This is supported by all those considerations yet to be given, which show that the dark spots are solids, and less luminous than the surrounding liquids.

14. The fixed stars have smaller bodies revolving around them, like the planets around the sun. On account of their vast distance only a few of the planets of the fixed stars have yet been discovered. Of the system of Zeta Cancri, it is affirmed that "the nearer companion appears

to have a motion ten times more rapid than the remoter one."* The following large stars, all of the first magnitude, have minute companions or planetary attendants of the magnitudes numerically annexed:

Sirius,..............10	Rigel,.............. 9
Procyon..............12	Antares,............ 7
Aldebaran,............12	

In all cases the minute companions have not yet been observed to revolve around the principal star, though they have around some. But this may be accounted for partly by the want of sufficient observation, and partly by the extreme lightness of many stars, and consequently their feeble gravitation. If this gravitation be extremely feeble, the motions of their planets must be extremely slow, and very hard to be observed. There is reason to believe that great differences exist among the stars in the materials of which they are composed. In our solar system, for instance, one member, Mercury, is sixty times more dense than another, the first satellite of Jupiter. The fixed lines of the different stars indicate vast differences between their materials; and among bodies so far asunder, one may be of a nature several hundred times less dense than another, and consequently the companions of the lighter must move so slowly as to require ages for their revolutions to be detected. The common proper motion of the companions of a double star, and the high improbability of so many near juxtapositions by mere perspective, lead us to believe that they are really binary, and revolve around one another, even when no such revolution is apparent. Such juxtaposition by perspective, however, is possible; as in the case of Vega, a star of the first magnitude, and its minute companion of the eleventh magnitude. The parallax of Vega, and the

* Humboldt's Cosmos, vol. iii., p. 287. Bohn's Edition in English.

want of parallax in its companion only 43″ distant, prove that they have no physical connection.

15. The binary stars move around one another, or rather around their common centre of gravity, in ellipses, the same species of orbit described by the planets around the sun.

16. The force of gravity animates the fixed stars the same as the sun. This is proved by their elliptic motions.

17. The centrifugal force reigns among the fixed stars the same as in the solar system.

18. Matter in the fixed stars has the property of inertia the same as here. This is shown by their centrifugal force.

19. From these four items just mentioned we are certain that matter in the fixed stars is subject to all the laws of motion which are so curious and precise among the attendants of the sun.

20. Atomic attraction, and

21. Atomic repulsion, are as active among the particles of matter in the fixed stars as in our own system. Repulsion is necessary to hold in a state of vapor the elements which cause the fixed lines; and attraction is necessary for the existence of liquids and solids.

22. The force of chemical attraction and repulsion must also exist in the fixed stars. These properties must go along with the chemical elements. We cannot think of separating them.

23. The sun and the fixed stars have their own proper motions, carrying their attendants along with them. The velocity of the sun has been estimated at 422,000 miles per day,* and that of 61 Cygni at 160,000 miles per hour. This last velocity results from its proper motion of 5″ per annum, and a parallax of 0″.348.

* Herschel's Outlines.

All these points of agreement between the sun and the fixed stars will be at once acknowledged by all astronomers, except perhaps that of solids in the sun, and this also, I trust, when the evidences in its favor shall be given on another page. But without this, what a wonderful accumulation of proofs do we behold that the fixed stars are suns like our own! It is undeniable that they give similar amounts of light; that their masses of matter are similar in weight; that the chemical elements exist in them the same as in the sun; that they are sources of the same intense heat; that gravitation reigns there the same as here; that they have smaller bodies revolving around them in elliptic orbits; that they rotate on their axes; that they have spots on their sides; and that they are moving with inconceivable rapidity through space. These are all fundamental facts, and they prove conclusively that those twinkling luminaries are really suns. With a firm faith we may rely on this modern revelation that God has created suns without number. Sir William Herschel believed that at least eighteen millions are placed in the Milky Way alone. If there be so many suns, how many more planetary worlds must revolve around them! And how great must be the Being who originated and who animates all these!

SECTION IV.

PLANETS, SUNS, AND MOONS ARE BODIES OF THE SAME NATURE.

HAVING proved that our earth is a star like the planets, and that the fixed stars are suns, it remains now to prove that suns and planets are in all respects bodies of the same nature. The distinction between them is altogether arbitrary, and hastily assumed from the accidental peculiarity of our solar system. According to our old ideas, a sun is the centre of a system, a planet revolves around a sun, and a satellite around a planet. This broad distinction holds good in some systems, and, as a matter of convenience in speaking and writing, it serves a handy purpose readily to denote which body or bodies we mean. But we have no right on that account to assume that their natures are essentially different. When we look away to other suns and systems, we see that the distinction of sun, planet, and satellite fades away, and they merge by insensible gradations into each other. In many cases the bodies of the system are two or three, and of no perceptible difference in size. Then in hundreds or rather in thousands of other cases there is a difference in size, and this difference ranges by small degrees from the least discernible to the very greatest. Sirius is supposed by all astronomers to emit at least sixty times more light than our sun. Other estimates make the light of that star twice or thrice greater; and yet his companion is so small as to have escaped the searching gaze of the telescopes of the Herschels and of Ross. One was first seen by Clark, of Boston, and since then Goldschmidt, of Paris, has announced the discovery of five more. I have already mentioned the minute companions of Procyon, Rigel, Aldebaran, Arcturus, Antares, and the two companions of Zeta Cancri, the nearer of which appears to have a motion ten times more rapid than the remoter one. Such comparative motions occur in our own solar system. Thus among the fixed stars there are systems like our own, having one large and several smaller bodies, and from this extreme case at one end of the series we see other systems of bodies becoming gradually nearer and nearer in dimensions, until at length, at the other end of the series, there is a perfect equality of size. This plainly shows that, naturally, as far as size is concerned, there is no line of demarcation between sun and planet.

Neither is luminosity a ground of distinction between sun and planet. Among the fixed stars which we know to be suns, we see all degrees of luminosity, from a brightness or amount of light far exceeding our sun, down through all degrees to perfect darkness. Some stars have only a small opaque spot on their sides, and at each rotation we see only a small diminution of their light. In others, this diminution extends further, and by degrees we come to others that are perfectly dark at each rotation. Lastly, others have permanently lost their light, and they are improperly termed lost stars instead of opaque stars. Thus we learn, first, that stars may be luminous in one period of their history and dark in another; secondly, that their surfaces, by reason of large spots, may be partly luminous and partly dark, or imperfectly luminous; thirdly, that small stars, which move around larger companions, and which are generally termed planets, are in some systems luminous, and in others opaque. In our own system they have all become opaque, but, as already stated, in the systems of Sirius, Procyon, Antares, Zeta Cancri, and very many others, they are still bright. Luminosity depends on the period in the history of the star, and not upon its nature. We will soon see that our earth, in the earlier part of its history, like other fixed stars, was self-luminous, and at a later period it became, like many other fixed stars, opaque.

Neither is position or movement an indication that there is an essential difference between suns and planets. Strictly speaking, a planet does not revolve around a sun as a centre, but both sun and planet revolve around their common centre of gravity. This centre is sometimes within the surface of the sun, sometimes a little without, then still further without, and so on by degrees, further and further, until both companions are of the same size, and then the centre of gravity is precisely midway between the two, and their motions are the same; then all distinction between sun and planet is completely lost. Thus, in size, position, motion, and luminosity, we find not the least particle of proof that the natures of sun and planet are different. On the contrary, by all these weighty attributes, they appear plainly to be bodies of the same general class.

Again, we find that in all these important particulars there is the same want of distinction between planet and satellite. Alpha Andromedæ, Mu Bootis, and Mu Lupi are double stars, consisting of a larger and a smaller star. In popular language the larger must be called a sun and the smaller a planet. But, has the planet a satellite? The remarkable case is this: With the most powerful instruments this planet is seen to be double—to consist of two individuals, the one a planet and the other a satellite; and their sizes are such as to show that there is no difference between planet and satellite. Eta Lyræ is seen as a double star; but, when viewed with more powerful instruments, each of these companions are again seen to be double. In this system there is no distinction in the first division between sun and planet, and in the second division there are no distinctions between planet and satellite. Xi Cancri, 12 Lyncis, and 11 Monocerotis are clearly double stars with a small distant attendant revolving around both. Mr. Lassell reports a star, which, with a high magnifying power, appears to be triple; the three companions are so placed as to form a right-angled triangle; and each companion has an exceedingly minute near attendant, thus forming a singular group of six.* Theta Orionis, when duly magnified, is seen to consist of four stars, not precisely on the corners of a rectangle, and hence the group is called the trapezium of Orion; and, at least, two of these stars

* Monthly Notices of the Royal Astronomical Society, vol. xxiii., p. 196.

have each a very minute companion. Such systems as these set at defiance the application of our familiar classification of sun, planet, and satellite, and show us clearly that our classification is arbitrary and not founded on essential and fundamental distinctions. By these examples we see that a satellite may be self-luminous, and that by position, size, and movement, it cannot be distinguished, in some cases, from a planet. Even in our own system, size is no ground of distinction; for one of the satellites of Jupiter, and another of Saturn, are larger than Mercury, and about the size of Mars, to say nothing of the asteroids.

We have now reviewed the evidences that the earth is a star like the planets, and that the planets are worlds like our own; that the fixed stars are suns, and that suns and planets are bodies of the same nature, having no marks of distinction between them except such as are arbitrary and applicable only to accidental peculiarities in systems like ours. All these bodies, whatever be their names, are composed of matter having the same essential properties. We behold in them all those modifications of matter which are denominated the simple chemical elements, and, therefore, we conclude that the chemical force of attraction and repulsion is active in all the great globes of space. Gravitation animates them all, showing its power inversely as the squares of the distances. Solids, liquids, and vapors exist generally among the stars, and therefore the atomic forces of attraction and repulsion must be equally prevalent. They all move in elliptic orbits, and therefore centrifugal force, inertia, and all the ordinary laws of motion, are as common among the fixed stars as on the earth. Nothing more is necessary to show that on our earth we are really treading on a fixed star. Here we have an opportunity of leisurely observing how a fixed star appears after its light has gone out. And when we calmly survey the bright gems in the nightly sky, we can behold how our own globe once appeared. As in a forest we note the progress of the oaks from the acorn to the tall tree, some just rising from the ground, others vigorous in the sapling growth, and others whose trunks are populated with mosses and lichens, and whose branches are alive with birds, so we can see like stages of progress among the heavenly bodies, our earth included. Some are glowing with the fervor of most intense heat; others, like our earth, are cooled on their surfaces, and with only volcanoes to tell of their molten interiors; and others, like our moon, are still further on in their history, where even volcanic energy has become cold and dead. Some are invariably bright, and others, like our sun, exhibit comparatively small, dark spots on their sides. Some at each rotation have their light slightly dimmed with spots, others again are dimmed more and more, and still others have at each rotation their light entirely hid. At last we behold others whose light goes out entirely, perhaps to be rekindled again by a temporary glow, and to be called by astronomers a "temporary star," and then its light is gone, dark forever—dark, not to be a dreary solitude, but in a resurrection morn to be reillumined like our earth, with the happy light of intellectual life and social enjoyment.

SECTION V.

EVIDENCES THAT THE EARTH WAS ONCE SELF-LUMINOUS LIKE THE SUN.

We have already seen that in some systems the planets are self-luminous, while in ours they are opaque. We have also seen that some self-luminous stars lose their light, and thus become dark like our earth. These two facts strongly impress the idea that our earth also at some former period lost its self-luminous power. We now enter upon the work of examining our earth as a fixed star, and of gathering all the proofs we can find in its constitution that formerly it shone as brightly as the sun. Geology has taught us many wonders of the structure of our globe and of its ancient history during an inconceivable number of centuries. We have read these wonders chiefly in relation to ourselves and to the millions of different species of plants and animals that from time to time have flourished on its surface. But now we will regard all geological facts as so many precious items in the structure and history of a star. We will study astronomy in the ground beneath our feet. And then having an intimate and a thorough knowledge of one star, we shall be the better prepared to study the others. As those far-distant orbs shed their radiance from all sides on our planet, so we shall find that intellectual light may be radiated from our earth to the fixed stars, and that with all their brightness they may receive a new glory from our researches in the constitution of our globe.

In proving that our earth once burned as brightly as the sun, we will review, first, those facts which prove that the interior of our globe is now so highly heated as to be in a melted condition; and secondly, those facts which prove that formerly both the interior and the present solid outer crust were one molten, flaming mass.

1. Deep mines in all countries, and on every side of the globe, furnish strong and unmistakable evidences of the high temperature of the interior of our planet. The ground near the surface is warmer in summer and colder in winter, but after descending a few feet, the temperature is invariable all the year, and this temperature increases the deeper the mines are sunk. When the mines are very

deep, the heat is more intense than the hottest summer weather. This is the case not only with the air and the rocks, but with the waters gushing on every side from the rocks and coming from the distant underground regions. This is proof that the heat cannot by any means be produced in the mine itself, but that it is the natural condition of the inside of our globe.

2. Artesian wells agree perfectly with deep mines. The deeper they are bored the warmer are their waters. If this were true in a few countries only we might suppose the heat to be due to some local causes; but it is true all around the globe, and hence we must believe that the interior of our planet is very highly heated. The average increase of heat in a downward direction, both in mines and in artesian wells, is one degree for about fifty-five feet. We can think of nothing to stop this regular increase of heat downwardly; and according to this ordinary rate, all the rocks must be so highly heated as to be in a melted condition at the depth of about thirty miles. It has been found that rocks when under extremely heavy pressure, require a much higher temperature for their melting point; and it has been rashly concluded that on this account the solid condition of the rocks must extend further down than thirty miles. But we know that the freezing-point of water may be lower than the melting-point of ice. Ice melts at 32°, but if water be kept perfectly still, its temperature may be considerably lower before it solidifies into ice. In like manner the rocks under heavy pressure may require a high temperature for melting, but when already in a fluid state their temperature may, for aught we know, be considerably lower before they solidify. To turn a solid rock into a liquid, is one thing; to keep an immense quiet liquid from becoming a solid, is another. Therefore, on the supposition that our globe was once in a fluid state,

we need not conclude that it must solidify at the high temperature requisite for melting rocks under extreme pressures.

3. Hot springs, sometimes only a little above the mean temperature of the ground, and sometimes heated to the boiling-point, rise up from the deep interior all around our planet. They add a powerful corroboration to the truths told by deep mines and artesian wells.

4. Volcanoes enlarge still further our ideas of the interior heat of our planet. There are more than three hundred active volcanoes, and the number of extinct volcanoes is much larger. But as some active volcanoes have been perfectly quiet and cold several centuries between two eruptions, so some volcanoes now regarded as extinct may again burst forth. And besides these, there are in all countries huge masses of trap and basalt rock which geologists believe to have been submarine volcanoes. Looking, therefore, at the three hundred active volcanoes, at the still larger number of extinct volcanoes, and at the trap and basalt as having been submarine volcanoes, we see wonderful evidences of interior heat. Moreover, these active volcanoes afford evidences that at vast distances, hundreds of miles apart, they are connected together by a subterranean ocean of melted rock. The vast amounts of melted material they cast forth are also in harmony with this idea. The streams of lava from the eruption of Skaptar Jokul in Iceland, in 1783, were fifty miles in one direction and forty in another; in some places they were twelve miles wide, and in narrow glens they were five hundred or six hundred feet deep. In the Sandwich Islands, in quite recent times, jets of lava one hundred feet in diameter and four hundred feet high have been cast up as continued fountains for weeks in succession. The huge volcanic cones, two or three miles high and thirty miles in diameter, all erupted

from the interior of the globe, tell the same tale. The intensity of this interior heat of our planet is illustrated by the brightness of the lava, which is often declared to be of dazzling whiteness, like the sun.

5. Earthquakes are always connected with volcanic eruptions, and they are especially frequent and violent in all volcanic countries. These two facts show their connection with the interior heat of our globe; and as earthquakes occur in all countries, they prove the operations of interior heat everywhere. Single earthquake shocks run as immense waves through very large portions of the earth's crust. The great Lisbon earthquake of 1755 shook northern Africa, all Europe, the Northern Atlantic Ocean, the West Indies, the North American continent, as far west as Lake Ontario, and how much further we cannot tell. We can understand how so wide a shock is possible if the crust of the earth floats like ice on an ocean of fluid rock beneath; but if we regard the globe as all solid, we cannot understand how so wide a shock can occur.

6. The well-known density of our planet lends powerful support to all these other evidences in favor of an extremely high interior temperature. All substances, whether liquid or solid, may be pressed by great force to occupy a smaller volume. The interior of our earth is under pressure exceedingly great. Even its outer surface is pressed downwardly by the atmosphere at the rate of about one ton to every square foot. What, then, must be the pressure a hundred or a thousand miles down by all the weight of the solid and liquid rocks! Such a pressure should make the globe many times more dense than it now is; and nothing imparts to my mind so impressive an idea of the great intensity of the interior heat as the fact that it can counteract all this inconceivable pressure and keep our globe at its present lightness. Its interior must be composed of the

heavier sort of its elements, such as the slightly oxidized iron, copper, lead, zinc, and other métals that do not oxidize, as platinum and gold. This we know, because such heavy materials must naturally gravitate toward the centre. And yet, heavy as the interior should be, both on account of these materials and the extraordinary pressure, the heat is so intense as to keep our planet comparatively light.

7. The unaltered length of the days, strange to say, is a strong proof of the interior heat of the globe. The tides of the ocean, caused by the sun and moon, are constantly striking against the eastern sides of all continents and islands, and forcing them with the entire globe westwardly. Even those tide-waves which strike the western shores are only *rebounds*, after spending their force in the opposite direction. This constant force of the tides, huge cyclopean strokes day and night for hundreds of years, should have a perceptible influence in causing the globe to rotate more slowly, and hence the days should grow longer. But for 2,000 years there has been no lengthening of the days, even by the smallest portion of a second. Hence there must be something to counteract this tidal influence. This counteracting force is the interior heat of the earth. It is slowly radiating away from hundreds of volcanoes and thousands of hot springs, and even through the ocean and the dry land. This loss of heat causes a contraction of the size of the earth, and this contraction tends to add velocity to its rotation from west to east. This tendency to an increased velocity of rotation just balances the retarding power of the tides. Thus the even balance between these two opposing forces keeps the rotation of the earth and the length of the days unaltered.

When the interior heat shall be all radiated away, or so much radiated as to cause a slower rate of contraction, then the tidal influence will be predominant, and the days will grow longer. They will continue to grow longer until there will be only thirteen days in a year—each day a lunar month, like a day now on the moon. And as the same side of the moon is now always turned toward the earth, so the same side of the earth will always be turned toward the moon. Which will be that favored side of our globe, no one can tell; nor can we tell how many hundreds of millions of years will pass before the coming of that far-distant event.

8. *The igneous rocks.* The entire solid crust of our globe consists of two classes of rocks, the igneous and the aqueous. Their origin is deeply interesting, because just as the interior is now in a melted condition, so the origin of these two classes of rocks shows that the exterior also was once in a melted condition. The igneous rocks are of various kinds, such as lava, trap, granite, and the metallic ores; but the chief one of these, forming ninety-nine hundredths of them all, is granite. The evidences that they have cooled and hardened from a former melted condition are these: Their position is below and next to the region of interior heat. On these rocks, therefore, rest all the other rocks; though in various places they are uncovered and exposed to view. They have filled up irregular cracks and cavities in the other rocks, the same as a casting fills a mould. They have produced all the evidences of heat on the other rocks, turning bituminous coal into anthracite, limestone into marble, clay slate into mica slate, and the ordinary red sandstone into quartz with dark carbonaceous markings. They are crystalline in their structure, in such a way as to show a previous liquid state. These crystals of granite are of three or four kinds—quartz, felspar, mica, and hornblende; and these could not have been formed and placed in their present positions out of a solid. They must have been formed from a liquid—a liquid from fusion by heat. Ordinary lavas from volcanoes, and trap, and granite, pass into each other by insensible gradations.

Often it is impossible to distinguish lava from trap, and trap from some of the forms of granite. Lava may cool very suddenly, and then it is not crystallized; if cooled slowly, it is crystallized, but the grains are small; trap has cooled more slowly, and the grains are larger; granite has cooled extremely slowly, and the grains of crystals are very large. As granite is so closely related to lava, and as we know that lava has been melted, so we conclude that granite has been melted also. From all these sources of evidence, no geologists doubt that all the igneous rocks have cooled and hardened from a former melted condition.

9. The Aqueous Rocks. These have all been deposited as sediment in the bottom of waters, and hardened by various cements. The materials of which they are composed, except the carbon, have been derived from the igneous rocks, and consequently they were at a former period, as we have just seen, in a melted condition. Their sedimentary origin is evident in many ways. They contain the remains of plants and animals, just as such remains are now embedded in the mud. They even present the footprints of animals, and the round marks of rain-drops, indicating by their fall the direction of the winds. Some are composed of coarse and others of fine pebbles, the same as if gravel-beds of different kinds were hardened now. The several varieties of sandstones and clay slates are the same as if our present beds of sand and clay were hardened into rock. It is highly instructive to look at the ordinary red building sandstone, and see the grains of quartz, feldspar, and mica. These show that they have been formed out of granite. Many granites partly decompose by the action of water on their lime, potash, soda, and magnesia, and then their grains not decomposed, become cemented again. The uncemented grains of quartz form the sands on the sea shore, and the lime, soda, potash, magnesia, and alumina,

have sought various new combinations. The lime has mostly united with the carbonic acid of the air and of the water, forming carbonate of lime; and this has supplied the material for bones, shells, and corals; and these in their turn have been deposited as chalk and limestone, and converted into marble. The beds of coal, of asphaltum, of oil, and of peat, are easily traceable to the carbonic acid of the atmosphere, which is absorbed by plants, decomposed, and the carbon secreted in various vegetable forms. The evidences are perfect that the aqueous rocks have been derived from the igneous rocks, and hence the entire crust of our globe has once been, like the present interior, in a melted condition. The only exception is the carbon and some other gasifiable materials which floated in the atmosphere of the ancient fiery world.

10. Trap Dykes. These consist of trap or basalt rocks rising up from the deep interior of the earth in the form of walls through the midst of whatever other rocks that may lie in the regions where they occur. They are generally in straight lines, from a few to many miles in length, and it is certain that they have been erupted from beneath in a melted condition up through long fissures in other rocks. The Palisades, on the west bank of the Hudson from Piermont to Staten Island, are examples; and the eastern mountains in New Jersey, extending from nearly the same northerly point as far southerly as Somerville, are similar parallel ranges. The peculiar form of these trap dykes, long, straight lines, a little roughened here and there, strongly favor the idea that the crust of the earth is comparatively thin, that it is subject to rents or fissures, that a bed of melted rock lies below, and that these melted materials may be forced up through the cracks. Their frequent occurrence in all quarters of the world adds powerfully to this view of the fused interior of the globe.

11 Faults. These are long fissures, often extending many miles in the crust of the earth, though generally no molten materials have been forced up; but the strata of rocks on one side of the fissures have been raised up or lowered down more than on the other side. This raising up or lowering down, has been in some cases a few inches, in others a few feet, in others a few hundred feet, and in others a few thousand feet. It is impossible that the crust of the earth should be raised up or lowered down, in this peculiar manner, through a great extent of country, unless there be an ocean of liquid rock below.

12. The Rising and Falling of the Earth's Crust. Various regions of the earth's crust are now known to be rising up or settling down. The most celebrated of these are the western coast of South America, Greenland, and Sweden. The Atlantic coast of North America south of Boston is being very slowly depressed. These changes of level cannot be caused by slow expansion and contraction in the rocks below from gradual changes in the subterranean temperature, because some of these elevations of surface, as on the west coast of South America, have been sudden, several feet at once for a hundred miles or more. Neither can they be caused by the sudden formation of subterranean gases, for these elevations remain permanent. The whole continent, with the mighty Andes, is in the slow process of elevation, as is proved by the raised sea-beaches and other marine objects hundreds of feet high. All the present elevations all around the globe are but continuations of former elevations which have lifted above the waves the continents and islands and the cloud-capped mountains. The floors of the oceans are slowly going down. The coral islands began to be formed with their present bases near the surface of the water. Now these bases have sunk down several thousand feet; the sounding lines have failed

as yet to tell how many thousand. But the coral animals have kept building up the tops of these islands as fast as their foundations have sunk down; and now their sides present an almost perpendicular wall. The greatest elevation of land is a little more than five miles, and the greatest depression of the ocean floor is supposed to be about seven miles. Therefore, the vertical movements in both directions, added together, have amounted to twelve miles. The great risings and fallings of the earth's crust presuppose a liquid interior to make such risings and fallings possible.

13. The Oblateness of the Earth. The polar diameter of our planet is about twenty-six miles shorter than its equatorial diameter, and this is the precise amount due to the earth's density and velocity of rotation. The peculiar shape of our globe, therefore, is exactly such as it must have acquired in cooling from a former liquid condition.

14. Lateral Pressure. Everywhere there are evidences among the rocky strata of great pressure in a horizontal direction: it may be called lateral pressure, for distinction from the vertical pressure. If we place on a table a quire or two of paper, somewhat dampened to make it flexible, and then press inwardly against the edges of the sheets, they will bend in waves up and down. Such appearances are common, especially among gneiss rocks. As this sort of pressure occurs all around the globe, there must be some widely pervading cause, and the discovery of that cause lends a powerful proof of the cooling and hardening of our globe from a former melted condition. All liquids expand and contract by changes of temperature much more than solids; thus the liquid mercury in the thermometer goes up and down because it expands and contracts so much more than the solid glass. After a crust had been hardened on the outside of our globe, it was like the bulb of a thermom-

eter. By a further and a continued cooling, the liquid interior contracted more than the solid exterior. The crust was then TOO LARGE! But it must accommodate itself to the shrinking fluid, and hence lateral pressure. It must squeeze up into a smaller circle, and like the dampened paper on the table it must rise and fall in waves. In the entire history of the physical creation of our globe, there is not a more prominent fact than this; none that has been attended with more striking consequences of various kinds, none that proves more strongly the former fluidity of our planet.

15. Dry Land. In the beginning there was no dry land; all the surface of our planet was water. This is proved by the marine character of all the fossil plants and animals in the earlier strata. And God said, Let the waters be gathered together in one place, and let the dry land appear; and it was so. This was done simply by cooling and by the contraction of the liquid interior more than the solid crust. That crust by lateral pressure was waved, rising up in some regions and settling down in others. These risings formed the continents and islands, and the settlings down formed the ocean floors.

16. Mountain Chains. These also bear witness to the former fluidity of our planet, and to its gradual cooling and contraction. They are but the upturned edges of a fissure in the crust of the earth. These fissures, as we have already seen in the trap dykes and faults, are in general long, and not far from a right line. When by lateral pressure a wave in the crust of the earth is bent upward, that wave in such rigid materials is liable to become a crack, a very long break in the solid envelope. By long-continued lateral pressure the edges of that fissure may be forced over each other, and become turned up and thrown in confusion. This violent disruption and elevation is a mountain chain. It

speaks eloquently of its origin and the past fiery history of our globe.

17. The prevalence of Hot Springs in Mountain Chains. Although mountains are always cold and often covered with eternal snows, even in equatorial regions, yet they are the favorite homes of hot springs. This anomaly may be solved by the consideration that mountains are but the upturned edges of fissures in the earth's solid crust. Through these fissures the interior heat of our globe has an opportunity to rise toward the surface. Hence the waters heated in these crevices appear soon on the surface as thermal springs.

18. Volcanoes in Mountain Chains. All around the globe the special seats of volcanic action are in mountain chains. They rise up from one end of the American continent to the other, through the great ranges of the Andes and the Rocky Mountains. The backbone ranges of long islands, as Sumatra, Java, and Japan, are other instances. But all these should be the very last places to look for volcanoes, if mountains were simply thicker portions of the earth's crust. If, on the other hand, we take the views of mountains just given, and regard them as the upturned edges of fissures in the crust of the globe, we can at once understand how the heated molten interior should find a most easy exit in such localities, openings ready made where it can burst forth.

19. The Parallelism of Mountain Ranges. It is quite common for mountain ranges to be not single, but for two, three, or more ranges to run side by side, with long, beautiful valleys between. In Pennsylvania the ranges seem like so many ocean waves. It is more or less the same all over the world. This is the natural result of lateral pressure; the same as in our paper on the table while pressed

edgewise, there appears not only one, but many ridges and furrows running side by side.

20. Mountain Chains through the centres of Long Peninsulas and Long Islands. These tell the same story of lateral pressure from the more rapid contraction of the fluid interior of the globe. Such mountains, like huge backbones, are conspicuous through the following long islands and peninsulas: California, Cuba, Scandinavia, Italy, Sardinia, Madagascar, Sumatra, Java, New Guinea, Niphon, the Malayan peninsula, and many others. All the islands and peninsulas are simply the elevated curves in the waves of the earth's crust. Lateral pressure forced them upward, and a break occurred in or near their centres, just where on mechanical principles a break was most likely. The edges of that break have been forced against and upon each other in wild disorder, as we behold in mountain crags. It is easy to understand in such cases how the mountain chains and the coast lines must be in the same direction. These islands and peninsulas are waves, and the mountains are the broken crests on those waves.

21. The Parallelism in Continents between their Mountain Chains and their Coast Lines. This is a remarkable feature in the great framework of the globe, and it leads directly to the cooling of our planet, from a former state of fluidity. The Andes and the Rocky Mountains hold the same directions as the Pacific coast, and the Appalachian range in North America and the Brazilian Andes in South America conform to the neighboring Atlantic coast. A mountain range also conforms to the northern coast of South America. The Atlas mountains in Africa run parallel to the shore of the Mediterranean; and another great range, recently discovered, runs from Abyssinia to the Cape of Good Hope, parallel to the shore of the Indian Ocean. In Europe we have seen a similar state of facts in

its large peninsulas. In Australia the chief mountain range runs parallel to the eastern coast, at no great distance. In Asia we have seen similar facts in its great peninsulas. In Hindostan the two Ghaut mountains run parallel to the eastern and western coasts. If we look at the map of Asia, we will see that several nearly parallel mountain chains extend from end to end in the direction of the greatest length through that continent, beginning near Behring's Straits and ending in Asia Minor. Regarding Europe as but the continuation of Asia, we see the general continuation of the central Asiatic mountain elevations through central Europe.

There must be some world-wide cause for the general recurrence of such great phenomena, and lateral pressure so often repeated is such a world-wide cause. Continents are only large islands, and the very cause which elevates a mountain lengthwise through the middle of an island or peninsula must do the same through a continent. Only the continents, being larger, must have a larger number of waves of elevation, and hence in the two Americas a mountain range faces each ocean. Hindostan, being a broad peninsula, has a mountain facing each coast. The broad peninsula of Spain has several parallel ranges in the direction of its greatest length. The broad continent of Asia has several parallel ranges lengthwise through its centre. As the entire liquid interior of the globe contracts, so the entire solid exterior must be furrowed with waves, and there must be much longer and higher waves in continents than in narrow islands. These waves must run parallel with the trends of the continental shores, for the same reason that they run parallel to the insular and peninsular shores. But our globe being round, there must be some contraction of continents in every direction. Hence the Ural and Beloor mountains in Asia are nearly at right angles to the general mountain trends of that continent. They represent the contraction of Asia in length. The same fact is represented in the American continent, by the downward bend in the central part, cutting it almost in two, and by the upward bend from east to west along the northern shore of South America.

22. The Position of Mountains on High Table-lands. Ordinarily, when we hear of the heights of mountains, we do not hear of their heights above the surrounding plains at their bases. These bases are generally elevated above the ocean level as high as the mountain tops are elevated above their bases. The table-lands of the Rocky mountains are 8,000 feet above the level of the sea. How has it happened, in the creation of the world, that lofty mountain chains are perched up on lofty table-lands? The reason is to be sought in the cooling of the liquid interior of our planet. This causes lateral pressure and waves in the solid crust. The same lateral force which raises up the mountain, raises up the high table-land. The table-land is the top of the wave on which the mountain is the crest.

23. The Greatness of the Features on the Earth's Surface. The chain of the Andes and Rocky Mountains, with frequent volcanic eruptions, extends well on to 9,000 miles. The whole American continent is only a single feature of our globe. The high table-lands and the associated mountain chains running through the Asiatic continent, with volcanic and earthquake forces in operation through the entire range, are but a single feature, and must be referred to a single cause. The Atlantic Ocean is the single downward curve of a wave between America on one side and Europe and Africa on the other. The Australian and the Polynesian field of islands reaches from eastern Asia nearly across the Pacific Ocean. To produce such comprehensive, such world-wide features on the outside of our planet, there must be a cause

in operation equally comprehensive and grand. We behold a cause precisely of this character in the shrinking of the fluid interior and the consequent lateral pressure and furrowing of the exterior. This is a universal cause, and this corresponds precisely with the world-embracing features under our review. As lateral pressure operates at once around the entire globe, it can raise its waves measuring thousands of miles in length.

24. The former High Temperature of the Atmosphere. The fossil remains of plants and animals that lived in ancient geological periods, indicate a tropical climate in all latitudes. Even in high northern latitudes the trees showed no marks of annual growth, which are produced by the alternations of winter and summer. This tropical climate everywhere was probably produced by the interior heat of our globe, whose crust has now become too thick to allow so abundant an escape of heat out in the air. It has been supposed that all the lands in these old periods might have been located in tropical or sub-tropical regions, and that such a location would have produced a universally warm temperature. But this supposition is inadmissible, because the land plants and animals, and those living in the sea, near the shore, in those ancient geological times, have been found in high northern latitudes, proving that the land was located then very generally as now; though, probably, the amount of dry land then was not so great as now.

25. The Causes of Earthquakes and Volcanoes. The heat of volcanoes, their favorite seats on mountain chains, and the connection between volcanoes and earthquakes, we have regarded as evidences of the interior heat of our globe. But the causes of earthquakes and volcanoes should be regarded as a distinct argument. From the contraction of the fluid interior of our globe, and the consequent lateral pressure and furrowing of the exterior crust, the

upward curves of that crust must be raised above the hydrostatic level of the fluid interior, and the downward curves must be pressed below that level. The great weight of the upward curves must tend to press the downward curves still farther below that level; and hence the fluid interior must be forced up through any openings it may find in any curve below the natural hydrostatic level. But as the downward curves of the solid crust may force the fluid rock above its natural hydrostatic level in the upward curves, so there may be outbursts of this fluid rock in the form of lava, far above its natural hydrostatic level—even in mountain tops. This forcing and jetting upward of the fluid rock beneath, must take place especially when by a movement of the crust a readjustment is made between the solid above and the ever-contracting fluid below. That adjustment in so thick and rigid a crust cannot always be made gradually. A sudden cracking and breaking must, now and then, be suffered. These sudden commotions are earthquake shocks, frequently attended with terrific volcanic eruptions. The melted rock or lava, by these sudden lowerings of the crust, is forced up through in the air, not always near where the settlings of the crust occur, but perhaps hundreds or thousands of miles distant, perhaps on the very opposite side of the globe. In this way we can account for such great overflows of lava as that of Skaptar Jokul, in Iceland, and those of the Sandwich Islands. In this way also we can account for the great masses of some volcanic mountains, as those of Ætna, Cotopaxi, and Chimborazo. They have left no hollow space down below equalling their own size, because the fluid interior fills up the vacancy. Besides this general cause for earthquakes and volcanoes, there may be other minor local causes from the effects of the interior heat on gaseous substances, like carbonate of lime, and on the water that may find its way down through fissures. Hence immense volumes of steam and gases sometimes explode, with deafening noise, carrying rocks and ashes high up in the air.

26. **The Geological History of the Dry Land.** This history shows that in the beginning the solid crust was of the same size as the fluid interior, and consequently even and level all around the globe; that soon the fluid interior began a process of contraction far more rapid than that of the solid exterior; that a furrowing of this solid crust also then began; and that this contraction of the fluid interior and this furrowing of the solid exterior, have since been going on continually. The process is active in our own day. It is wonderful how all the great leading facts of geology support this view. That the exterior crust was at first level, and perfectly conformable to the liquid surface, is proved by the fact that then there was no dry land. No furrowings of the solid exterior raised themselves above the ocean level; and this state of things continued during the Cambrian and Silurian periods. But the furrowings proceeded so far as first to show their tops above the waters during the Old Red Sandstone and Coal periods. The forms of the early furrowings were perfectly characteristic of our theory; for the dry land, during the coal period, consisted of very low, level, moist islands. A still further furrowing, that is, a greater depression of their downward curves, and a greater elevation of their upward curves, put an end to the Coal period. Mountains of moderate elevation then first began, such as our Appalachian range. But as the process of cooling, contracting, and furrowing went on, the surface of our globe became more and more diversified in hills and mountains. Hence new tribes of animals and plants were from time to time created, and adapted to these new conditions. At length, much higher mountains were raised up, as the Andes, the Rocky Mountains, the Alps, and the Himalayas.

These are all new mountains in the world's history, because their fossil remains show that they are formed of strata which were still below the waves during the later Secondary and Tertiary periods. These elevations of the upward curves and the depression in the downward curves are still going on; because all round the globe, raised-up sea-beaches are seen from point to point, at higher and higher levels above the present ocean shores. Shells, of species now living, are found entombed in strata many hundred feet high. We may here recall to mind what has already been said, a few pages back, about the risings and fallings of the earth's crust in our own day.

The bearings of all these facts are very strong in our argument. They are exactly such as are necessary on the supposition that our globe was once a fiery, liquid mass, with its fires just burned out. Its heat must then radiate away, a solid shell must harden on its surface, the fluid interior must contract more than the solid exterior, that exterior must be pressed into furrows, an elevation must rise here, and a depression must sink there. As time rolls on, and the heat radiates more and more, the furrows become greater. The surface of our globe becomes more and more diversified; and in accordance with these ever-advancing conditions, more diversified forms of plants and animals are called into being.

From this survey of the history of the dry land it seems, at first view, that when a continent or a large island begins to rise above the ocean level, it should keep on rising steadily, and always grow broader and higher. But there are facts enough to prove that during this general rise there have all along been occasional depressions of the land with reference to the ocean level; that some of these depressions have been universal around the globe, and that others have been local. Both these kinds of depressions must now be explained. We must bear in mind that our earth is a globe, and that its solid surface, whether above the water or under the water, is convex. In its general furrowing, some of its upturned curves may be beneath the ocean. Now, in contracting and increasing its furrows, some of these upturned curves beneath the ocean may be raised higher. This would be raising the ocean-bed, and causing the waters to overflow the dry land all around the globe, providing that this dry land had not also been raised at the same time. Hence there would appear to be a universal depression of the continents and the islands, when, in fact, none had occurred. The ocean level relatively has risen; the dry land has been stationary.

These elevations of the ocean and these apparent temporary depressions of the land have taken place, from time to time, since the earliest geological history. During the periods of depression, deposits of sediment take place on the submerged flanks of continents and islands. The deposition of sediment cannot well be permanent on a slowly rising coast, except in some few small favored localities, because as the bottom rises it becomes exposed to the grinding power of the waves and currents. These wear it and carry it away. Therefore, the several great geological formations since the Silurian, must have been deposited during these periods of apparent continental depression. The general process of continental formation is one of elevation. As we have seen, the land is rising higher and growing broader. The time altogether has been immense. But the periods for depression were comparatively short, occurring only now and then, during the long process of creation. If animals and plants have been slowly changing their forms from the first creation until now, we must not look for many evidences of those changes from the beginning to the ending of any one geological formation, because those periods have been comparatively short. We should look for the progress of these changes rather in the long intervals between the periods of apparent depression; or, in other words, between the seasons for the deposition of the several great formations on the submerged flanks of the continents and islands.

We will now attend to the local depressions of the earth's crust. Let the straight line represent the ocean 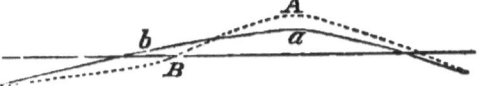 level and let the smooth curved line represent the contour of a continent. Then as cooling, and contracting, and furrowing goes on, the contour of the continent will be represented by the dotted curved line. At a, it has been elevated to A; but at b, it has been depressed to B. Hence, for a small distance, the continent, while generally rising, has been laid under water temporarily. Its deficiency of rise just there may have been caused by some weakening of the crust by interior heat or by fissures, and hence a depression. Or the bearings of the mechanical lateral pressure may have been such as to lower that part. In the great strain of this lateral pressure we can see how furrowing must occur; but we know too little of the bearing of part upon part in the unhomogeneous, irregular, and fissured crust, to say that every spot must keep on in a regular, continuous rate of rising; or to say that in this tremendous struggle there may not be local oscillations in a continent which is generally rising. We cannot say that the rounded back of a continent must always keep the same curve. The general curve may keep on, but it has irregularities; and these irregularities may not necessarily always remain the same. Hence, in a general rise, there may be a few local depressions.

27. The Larger Amount of Land in the Northern than in

the Southern Hemisphere. This fact is always prominently portrayed by geographers, but by geologists it has hitherto been passed by without any attempt at explanation. An explanation, however, may be perceived, if we regard our earth as a star, and attend to its astronomical relations during the period when the solid crust began to form over the fused interior. Such an explanation must confirm the theory that our globe was once in a fiery condition, like the sun in the heavens; and for this reason it must be introduced here.

The amount of heat the earth receives annually from the sun is very great; but during a day in that period when the earth is at or near its perihelion, it receives a larger quantity than the average, by one-fifteenth of the whole amount. This excess of heat is received at the present time on the southern hemisphere. As the axis of the earth makes an entire revolution in the direction toward which it points once in 25,868 years, therefore there is an alternation once during that period in the exposures of the two hemispheres to the sun during the season of perihelion. But we will suppose when the solid crust was first formed and still very thin, that the position of the axis was the same as now. It is true that both hemispheres receive during the entire year the same amount of heat; the excess of the southern in summer is equalled by its deficiency in winter. Still, during the season of perihelion, the southern hemisphere now receives more heat than is possible for the northern hemisphere to receive with the axis in its present position. It has a greater intensity of the sun's rays, as just stated, by the one-fifteenth of the whole amount. Now, what must have been the effect of this excess of heat on the furrowing of the thin crust? It is easy to understand that, in the contraction of the fluid interior, the weakest parts of the earth's crust must settle down.

Think of the thin, solid crust, as being too large to repose on the fluid interior. Its arch-like form will tend to hold it up; but the arch is too flat, and the crust too thin and weak to sustain itself, and therefore there must plainly be a caving-in somewhere; and that portion must cave in which is the weakest; and that must be the weakest which is exposed to the most intense heat. Even wrought boiler-iron is weakened and made more liable to fracture and explosion by heat. Once every year the yet tender crust of the southern hemisphere must be more exposed to heat, and, consequently, must be weaker than it is possible for the northern hemisphere to become. Therefore, in the adjustment in size of the solid crust to the contracting fluid interior, the southern hemisphere would be the first to cave in. In the great process of furrowing, that hemisphere would be in a downward curve. Lateral pressure would force it down still lower—even below the hydrostatic level. Therefore, the ocean would more extensively cover that part, and the great body of dry land would lie in the northern hemisphere. And thus we perceive that the great preponderance of land in the northern hemisphere is due to a former fused condition of our planet, and to its subsequent cooling in special astronomical relations.

28. The History of Animal and Vegetable Life on the Globe. This wonderful history agrees perfectly with a former fused condition of our planet, and a subsequent cooling, contracting, and furrowing. For the sake of brevity, this agreement will be summed up in the following four particulars:

Firstly. The history of life on our globe points to an early period when life here had a beginning; and, of course, to a previous period, when terrestrial life did not exist. As we review the forms of life, both vegetable and animal, from our own day backward through the long tracts of time, or rather of eternity, we perceive a gradual simplification of these forms the further we go back. From their present high organization there is a regular descent to a former very low and simple condition. This view brings us to a vanishing point of organized life. Although we see it not, yet we may form an estimate of where it is. Just as we may form a rude idea of the termination and height of a pyramid, by observing how it becomes less and less in an upward direction—even though its top be enveloped in clouds—so we may form a rough estimate of the period for the beginning of life on our globe, by observing how, in going backward, life becomes less and less complicated in organization, even though no remains of the first plants and animals be found. This period, for the beginning of terrestrial life, tallies exactly with that when a solid crust first cooled, and allowed a sufficiently low temperature for organized beings.

Secondly. The history of life on our globe shows that animals and vegetables began in the water. All the early fossil remains are marine. This agrees with the process of a cooling globe from a former melted state. At first, it could not have been furrowed enough to form dry land.

Thirdly. The animals and the vegetation of the Coal period show that then the land consisted of large, low, level, moist islands. The theory of a cooling globe requires that this must have been the very form and condition of the first land.

Fourthly. The history of life on our globe shows that organized beings were at first very simple in their forms, and that ever since then they have become more and more diversified, and adapted to new and more diversified conditions. This agrees with the law of cooling, contracting, and furrowing. The entire crust of the globe, both above and below the ocean level, is becoming more diversified in its conditions. The lands are being raised higher, new valleys are being formed, and the mountains are becoming more numerous and more elevated. This process of diversification has been going on since dry land first arose. Now, it is a great and significant fact, that a parallel process of diversification has been going on in animal and vegetable forms.

Thus, in looking briefly though broadly at the history of life in these four particulars, we see how it lends support to the view that our earth has cooled from a former melted condition.

We have just reviewed twenty-eight facts, or rather classes of facts, to prove that the interior of our globe is now so highly heated as to be in a melted condition, and that formerly the exterior also was bright and fiery like the sun. These many facts or classes of facts are all essential and fundamental. They embrace all the chief features of our globe, and thus all its leading features proclaim its gradual cooling from a primitive state of igneous fusion. No small matters are introduced here. All are vast, worldwide phenomena, and all bear with a powerful and comprehensive grasp on the argument. Let us pause and consider for a moment their world-embracing scope. They are these: Deep mines all around the globe increasing in heat downwardly at the rate of one degree for about fifty-five feet; artesian wells everywhere repeating the same truth; hot springs boiling up in all countries; hundreds of volcanoes, and situated on every side of our planet, sending up the liquid rock often as bright as the sun; earthquakes in every country, but especially connected with volcanoes, and thus proving their connection with the interior heat; the want of density according to the great pressure on the interior of our planet, showing the mighty repulsive power of the heat within ; the unaltered length of the days, indicating that the escape of the heat and the consequent contraction just balances the tidal influences from the sun and the moon ; the igneous rocks giving unmistakable evidence of having cooled from a former molten condition ; the aqueous rocks plainly derived from the igneous rocks, and thus showing that the entire solid crust has once been fused the same as the deep interior ; the trap dykes revealing former straight fissures many miles in length in the solid crust of the globe, with the liquid interior oozing through ; the faults, where the settling down of one side of a long, straight fissure, indicates a liquid interior to allow such depression ; the risings and fallings of the earth's crust, which, by their suddenness in many cases and by their permanence, presuppose a cooling and contracting interior ; the appearances of lateral pressure around the entire globe, and explainable by the law that the liquid interior in cooling must of necessity contract more rapidly than the solid exterior, thus producing lateral pressure and waves or furrows in that exterior ; the elevation of the dry land, especially in the period when it arose, not in the very beginning nor at a very late day in the world's geological history, but just when cooling, contracting, and furrowing, ought first to have brought it above the waters ; the mountain chains in long lines exhibiting the broken and upturned edges of fissures in the solid crust ; the prevalence of hot springs in cold and snow-capped mountains, confirming our convictions that mountains are fissures in the external envelope of our globe where the heat can most easily come forth ; the favorite homes of volcanoes in mountain chains, akin in their evidence with that of hot springs ; the parallelism of mountain ranges side by side like parallel ocean waves, just as lateral pressure actually waves or furrows a dampened quire of paper ; the mountain chains through the middle of long islands and peninsulas and also lengthwise through continents, precisely where lateral pressure should place them; the parallelism between mountain ranges and coast lines pointing out lateral pressure as the same cause for the origin of the mountain and coast line; the positions of mountains on high table-lands, like crests on the tops of waves, just where the fissures in the upward bent crust of the earth should occur, and where by lateral pressure the broken edges of that crust should be piled up in wild confusion; the greatness of the features of the globe, requiring an operative cause like contraction coextensive with the globe; the former high temperature of the atmosphere, most easily explainable by the interior heat coming through when the earth's crust was yet thin; the causes of earthquakes and volcanoes, plain and easy to be comprehended when we think how the solid crust must by jars and breaks accommodate itself to the shrinking fluid interior, pressing its downward curves in that fluid and causing it to rise; the geological history of the dry land, showing that one continued process of cooling, contracting, and furrowing has been going on from the beginning when the earth's crust was even and under water, to the present period of high continents crowned with higher mountains; the greater preponderance of dry land in the northern hemisphere, which accords with the circumstance that the first furrowing of the earth's crust took place when the southern hemisphere was turned toward the sun while in perihelion; the history of animal and vegetable life on the globe, proving that terrestrial life had its beginning at a definite period in the world's history, that that period occurred when the temperature was first low enough to admit of organization, that then the crust of the earth was not yet sufficiently furrowed to elevate the dry land, that the first land consisted of low, level, moist islands, precisely according to the law of cooling, contracting, and furrowing, and that since then both plants and animals have become all along more and more diversified in their organizations, in harmony with the greater and greater diversification of the earth's surface.

Undeniably this array of facts is extraordinary for their number, their greatness, and their direct bearing on the point. To all these there are no opposing facts. The whole world in all its departments, both in its past history and its present condition, speaks with one language, clearly and forcibly bringing us to the period when it was a fused luminous planet. While all these facts point backward to the very ancient period when the star on which we live was in a bright, fiery condition, we must regard this truth chiefly as an astronomical truth. Its astronomical bearing is its highest bearing. It adds one more to the catalogue of lost stars. As other stars have lost their light and ceased to twinkle in the blue sky, so we learn, by this examination of the star beneath our feet, that this also once shone brightly, and then arrived at a period when it lost its independent luminosity. As around other suns there are revolving planets, still shining with their own independent light, so we learn by these facts that our own planet was once as independent in light-giving power as they. But the great astronomical bearing of the present investigation is still to be unfolded. We may now learn what heated and lighted up our globe in star-like splendor. And knowing this, we are in the legitimate way for learning what lights up the sun and the other stars. We now see the value of the facts forming the introduction to this volume, and collected there in large numbers, to prove that our earth is a star; that essentially there is no distinction between fixed stars, suns, planets, and satellites; that all are fundamentally of the same nature, however they may differ in minor features. Therefore, if by an examination of one star we can learn the cause of its light and its heat, we have the very best foundation for learning the cause of the light and the heat of all the other stars.

SECTION VI.

PROOFS FROM CHEMISTRY THAT THE EARTH WAS FORMERLY SELF-LUMINOUS. THE CAUSE OF LIGHT AND HEAT OF THE EARTH AND OF THE OTHER STARS.

THE science of Chemistry teaches very clearly, from the earth's composition, that in the early history of our globe, it must necessarily have been on fire, first in a gaseous, and then in a liquid state. The argument is embraced in the following five propositions:

Proposition First. The elements which compose our earth are of such a nature, that their chemical combination must produce heat sufficient to melt, and even to vaporize or render gaseous, the entire globe. These elements are about sixty-three in number, and the possibility of their existing in a gaseous condition, is illustrated by the fact that if they were now uncombined at ordinary temperatures, about half of the globe would be in a gaseous state.

Proposition Second. The elements of our globe have a most powerful tendency to combine and burn. Their combustibility is one of the most beautiful and wonderful departments in the science of chemistry. Phosphorus burns quickly in the open air, and so does even lead if finely divided. Iron in the form of steel burns most brilliantly in oxygen. Pulverized antimony and some other metals burn vividly if thrown in the gas chlorine or the vapor of iodine. Potassium and sodium take fire if thrown on water. Silicon forms nearly one-fourth of the globe, and oxygen forms nearly one-half; and the two unite and burn spontaneously. The burning quality of sulphur is well known; and oxygen and hydrogen unite with the most intense of all heat, except that of electricity; the ashes or residue they leave is water. The common salt we eat is but the ashes of a most violent combustion between chlorine and sodium. Indeed, the first and most remarkable property of the elements of our planet is to combine and burn.

Proposition Third. The elements or simple materials composing our globe, are everywhere in a combined or burned condition; that is, they now exist as the ashes of a former great conflagration. As certainly as a heap of ashes convinces us of an amount of burning according to its size, just so certainly do the materials of our globe convince us of an amount of burning according to the size and weight of our planet.

Besides these evidences arising from the general combination of the elements, there are two special facts exceedingly impressive. One is, that some of the elements are only partially burned. As around a building burned by fire, we see here and there a half-burned or half-charred fragment, and portions of a wall still standing, so among the fuel of the great conflagration of the globe, we see here and there a portion half-burned, or scarcely burned at all. Much of the iron, for instance, is only partially consumed; instead of appearing as a peroxide, it is often only a sesquioxide. Copper is often nearly unscathed. Such materials have been partially protected from the flames, and we readily see how this has happened. They are naturally heavy, and have settled away from those light gaseous elements with which they would readily have combined. The lighter elements, as carbon, boron, silicon, potassium, sodium, calcium, aluminium, sulphur, phosphorus, and the like, are invariably combined, thoroughly burned, except where agencies are at work, as in trees, separating the carbon, and, in volcanoes, decomposing the sulphur compounds and setting the sulphur free.

The other special fact is that those elements, which would remain separated and uncombined in a conflagration, are now found really separated. They are very significant in their evidence. Among these are gold, platinum, and nitrogen. They combine with other elements according to well-known laws and by appropriate agencies, but a great conflagration would set them free from their combinations, and present them just as we find them. They are the standing walls and columns that will not burn; the same as some of the iron and of the copper, are the half-charred remains which have partially escaped from the conflagration. The case of our globe is quite as strong, if not stronger, than that of the burned building. The great facts that those elements are united which heat would unite, that those elements are only partially united which, under the circumstances, could partly be prevented from coming together, and that those elements are separated which combustion would separate, form an assemblage of evidences surprisingly strong. The longer a person studies practically the geology, the mineralogy, and the chemistry of the globe, in a genetic point of view, the clearer do the evidences become that our planet was once all aflame like the sun.

Proposition Fourth. The combination of the elements of our globe took place rapidly, and therefore with the evolution of the greatest amount of heat. This combination, from the nature of the case, could not occur slowly. More than half of the elements are solids, as all the metals, with silicon and carbon. No one can conceive how all these solids at the centre of the globe could slowly combine with the lighter gaseous materials at the surface. A partial combination of a few feet in thickness might ensue, and no more. The external coating of oxides would completely protect the interior from further oxidation. In the cases of zinc and some other metals, the thinnest conceivable film of oxide arrests the oxidizing process. In order to combine, the elements must originally have been in a gaseous condition, or part liquid and part gaseous—the very condition pointed out by the nebular theory as the primitive form of matter. In this state they must all have entered into combination, one with another, and they must have made our planet equally as bright and self-luminous as the planets we behold revolving around some of the distant fixed stars.

Proposition Fifth. We have considered the powerful tendency of the elements of our globe to combine and burn; we have seen that they could not combine slowly, but that they must have rushed together with the evolution of much heat and light, unless indeed they were created originally in a state of combination. I now endeavor to show, as my fifth proposition, that they were not created in combination, but that they were formed separately. This will agpear from three considerations:

1. The invariable process of creation is from the simple to the complex. Innumerable examples are around us, and a few of these will be enough for illustration. The lightning strikes down from the clouds, and the vapor of water in the air is decomposed into oxygen and hydrogen. Now mark what happens from these simple separate elements. It is the regular process of creation established by the Author of all things. The oxygen unites with the nitrogen of the air to form nitric acid. The hydrogen unites with other portions of the nitrogen to form ammonia. The nitric acid and the ammonia also unite to form nitrate of ammonia. This complex mineral is soon dissolved in the rain-drops, falls to the earth, then it is absorbed in the ground, then imbibed by the roots of plants, then passed into other combinations by the leaves and other organs of

the plants, then in different forms it appears in the fruits, and lastly, it is eaten by animals and men, thus aiding to form their bodies. Here we see the long process of creation between the lightning-stroke, the three simple elements —oxygen, hydrogen, and nitrogen—and the bodies of men, of animals, and of plants.

Men and animals breathe, and trees burn; and by both processes the oxygen of the air unites with the carbon of the wood and of our bodies, and thus forms the compound carbonic acid. This flies away in the wind, dives down in the water, and unites with the lime dissolved in the waters of springs which are brought up from the deep interior of the ground. This union forms the compound mineral, carbonate of lime. All carbonate of lime is formed in this way. In the igneous or primitive rocks there is lime in abundance, but no carbonate of lime. Here, therefore, as an historical fact, we see how the Deity creates a compound mineral out of the simpler materials. Now, observe the importance of this compound. It is called also limestone, marble, and chalk. It crystallizes in about three hundred different forms, many of which are exceedingly beautiful, and among these is the double-refracting Iceland spar. It chiefly makes up the hard parts of animals, as corals, shells, and bones. One of the first steps only in this progress we do not see now, because it was all done long ago; it is the union of oxygen and calcium to form lime; but we can separate them, and form lime artificially.

These five simple elements—oxygen, hydrogen, nitrogen, carbon, and calcium—are the principal ones that form the bodies of animals. They are first combined, as we have seen, into water, ammonia, nitric acid, carbonic acid, and lime. These five compounds are united together to form double compounds, as nitrate of ammonia and carbonate of lime. Then they are made yet more complex in vegetable forms; and lastly, more complex still in animal forms. But in converting the compound minerals first into vegetable and then into animal bodies, it is most astonishing to see how the Divine Architect proceeds from the simple to the complex. By appropriate agencies He forms little cells, visible only with the microscope; from these He forms seeds, and from the seeds He forms, ultimately, through a long routine, the tree loaded with its brilliant blossoms or rosy fruit, or the grain yellow for the harvest. From these fruits, and grains, and leaves, innumerable species of animals derive their bodies: first as simple microscopic cells, then as eggs, then as birds that warble in the branches, and lastly, as the crowning achievement of all—man,

"Whose heaven-erected face the smiles of love adorn."

If there be any truth more plain than all others, it is that God, in creating, upholding, and governing the world, works by agencies and according to law, and that His process is invariably from the simple to the complex: first the blade, and then the ear, and after that the full corn in the ear. First the mustard-seed—whatever in Oriental phrase that may be—the smallest of all seeds, and then the spreading tree, in whose branches lodge the fowls of the air. This chain of dependencies from the body of a man back through plants and compound minerals to the simple elements, is a large one, interlinking many agencies and laws; and it is most irrational to say, that in the original creation God began somewhere in the middle of this chain, say with compound minerals. No! He began with the beginning; this is "the way of the Lord;" He began with the simple elements, and combined them according to His ordained and combining laws! Those elements, combined according to those laws, must have produced heat enough to fuse the globe.

THE CHEMICAL FORCE. 63

For our present purpose it is enough to go back as far as those modifications of matter called the simple elements. But really we cannot say that even this was the beginning. What was before them, and how and when they were formed out of a more simple, homogeneous matter, will be another inquiry. But go back as far as we may, we can never hope to arrive at the real beginning in time, any more than we can hope to reach the bounds of space. Still, our plain duty is to inquire into the works of the Eternal and the Omnipresent as our abilities allow, farther and farther away, both in time and space.

2. The simple elements were formed separately, and afterwards combined; because God has created a special agency called "CHEMICAL FORCE," acting according to a complicated system of laws, whose object is to unite or combine these elements. The more we look at this wonderful agency, the more we must be convinced that by this means the Deity has put together the particles of matter, and by this He now holds them in combination. It exists through all known space—like the medium producing light, like the medium producing heat, like the medium producing gravitation. Its vibrations come from the sun, and far away from the fixed stars, producing chemical changes, and giving origin to the art of photography. Its universality is further confirmed by the existence of simple chemical elements in the sun and in the fixed stars. By the motions of the magnetic needle we have evidence of still another medium—the magnetic, reaching everywhere through the immensity of space. What relation exists between these five ethereal media causing light, heat, gravitation, magnetism, and chemical action, whether they be independent or merely different operations of the same thing, does not concern us now; we are concerned merely in the existence and the universal extension of the great agency

called chemical force, whose office is to unite the simple elements into compounds. As heat penetrates all bodies, as light penetrates the hardest diamond, as gravitation penetrates from the centre of the sun to the centre of the earth, as magnetism penetrates through all things, so the ethereal medium we call chemical force has its special home, its favorite residence, in the interior of all aggregations of matter.

The complicated system of laws by which this agency operates, may be expressed as follows:

a. It combines the simple elements in definite proportions, by weight, and these proportions are called their combining numbers.

b. It combines the simple elements sometimes in more proportions than one, and then the additional proportions are multiples of the first.

c. It combines compound bodies together, and then the combining number of any compound, however complex, consists of the sum of the combining numbers of its simple elements.

d. It combines gases in definite proportions by volume as well as by weight; but the combining numbers by volume and by weight are quite different.

In the illustration of these laws consists some of the grandest wonders of chemistry. When we thus look at the agency created for uniting the simple elements, when we see how it extends through all space, when we learn how it penetrates all solids and fluids, when we contemplate its system of laws according to which it unites the simple elements, nothing seems plainer than that here we behold the means by which the Deity has really combined all compound minerals as we find them. By this He now holds them together, and by this He originally put them together. Shall He create this to us infinite agency for combining the elements, and yet combine them in some other unknown way? Shall He ordain these complicated laws, and be the very first to overlook them or set them aside?

We must regard this chemical force, with its several laws, as a system of machinery extending through infinite space, and built by the wisdom of Omnipotence for one of the greatest conceivable material ends. All material things are formed by this force, and in accordance with these laws; every inch of ground beneath our feet, every drop of the ocean, and even the very gases of the atmosphere, are ready to leap into combination in precise and orderly obedience to this system of machinery. Beautiful, grand, and sublime, we cannot suppose it has been capriciously set aside, and some other unknown and inconceivable agency used for creating chemical compounds. Therefore the conclusion is most easy and natural, that the simple elements were originally created separate, and then combined by the wonderful medium now before us, called Chemical Force.

3. The simple chemical elements were originally created separate, and afterwards combined by laws, because their creation in a compound state would have involved an infinite number of miracles, without any object for such miracles. Above all things, the Author of the universe is a God of order and of law, especially when things are to be done on a large scale. If millions of grains of wheat are to be formed during a summer for the sustenance of man, He forms them all grandly by law. To form a compound mineral otherwise than by the laws above given, would be a miracle without any assignable object. It would be a miracle in mere idleness, the very thought of which is an impiety. And then how many millions of such miracles, all contrary to this sublime machinery and these beautiful laws, would be necessary! Take, for instance, oxygen alone. If created at first in a compound state, it would have to be

created in innumerable independent parts and conditions: one part in combination with iron in one proportion, another part with iron in a second proportion, and another in a third proportion; other quantities with sulphur, some in one proportion and some in another : and so on in independent bits, no one can tell how many of oxygen alone. What is true of oxygen is true of the other sixty simple elements in various degrees; and the number of strange objectless miracles that would be necessary on this plan no man can estimate.

But even this is a most superficial view of the number of these supposed miracles. In carbonate of lime the three elements calcium, carbon, and oxygen are not present simply in their due proportions taken as a whole ; but part of the oxygen is united with the carbon in one proportion, and part with the calcium in another proportion, and afterwards these two compounds are united to each other in due proportions. In this compound, therefore, the oxygen is in two very distinct parts. The same may be said of the elements in compounds forming nearly the entire crust of the earth. These compounds are generally silicates, and not only double compounds, but even triple, quadruple, and quintuple compounds. Silica, acting as the acid, is combined with several bases, as in mica, feldspar, hornblende, and many other allied minerals. In each of these millions of grains of minerals, the oxygen, if created in combination, must have been created in distinct combination with each acid and each base, in all cases in different proportions, but yet all the oxygen precisely alike all round the globe, and never varying the least in its peculiar properties. Now to suppose that oxygen, or any other simple element, ever was created in this way, appears to me the most monstrous and incredible of all suppositions. It seems next to lunacy.

From these three considerations we cannot avoid the

conclusion that the elements composing our globe were not created in combination. Every one must have been formed originally in a separate condition, and afterwards they must have combined and burned. How and when they were created, and what were the agencies employed in their creation, is another inquiry. That inquiry we will enter upon in another part of this volume, where I shall endeavor to show that they are merely modifications of an original primitive state of matter, and that they were formed during the long period of the condensation of the globe from an inconceivably rare gaseous state, commonly called the nebulous condition.

Our five propositions are now established: First. The elements that compose our earth are capable, if combined together, of heating the globe to a fused and even to a vaporous condition.

Second. These elements have the greatest conceivable tendency to combine and burn.

Third. These elements exist now, with small exceptions, in combination.

Fourth. This combination could not have taken place so gradually as to produce only a small amount of heat.

Fifth. These elements were not created originally in combination. Therefore they must have entered into combination in large quantities in a gaseous or liquid condition; and in so doing, our planet must have flamed as splendidly as the sun.

The former intensely heated condition of our globe, and the present state of combination of its simple chemical elements, are two distinct and acknowledged facts. They are two very different things, but yet each one is absolutely necessary for the existence of the other. We cannot see the possibility of the evident heat without the chemical action, and we cannot see the possibility of the chemical action without the heat. The two are correlatives of each other, and they are indissolubly united in time, place, and action. Here in the very constitution of our globe, in the identical place where we behold the evidences of heat, we also behold the materials all burned up which were sufficient to produce that heat. This cause for the heat is not far-fetched—it is right before us; wherever we see the signs of the heat, an amply adequate origin of that heat stares us in the face. We are not left to inquire whence came the fuel or how it was kindled, for we cannot avoid seeing the fuel, and we know perfectly well that its kindling is spontaneous. Chemical action on so grand a scale must have been attended with heat on a scale equally grand. The time for both was the same, in the forming state of our planet, when according to the nebular theory all was gaseous or liquid. In all this we are impressed with no obscurity, no mazy theory, nothing dim or hard to be understood. Because the facts of the case are so plain, simple, and directly on the surface, they have been passed by in search of something hidden and strange.

It is wonderful how many different departments of creation, how many of the sciences, prove that formerly the earth shone like the sun.

Astronomy teaches that the earth is a star; that in all fundamental particulars it is just like the other stars that glitter in the sky; that as other stars are known to lose their light, so it is not strange that the earth is no longer self-luminous; and that even now there are self-luminous planets which revolve around other and far-distant suns.

Geology teaches by more than twenty great classes of facts that the interior of the earth is still in a fused condition, and that formerly the exterior also was fused by burning heat.

Physics or Natural Philosophy teaches that the shape

of the earth is precisely such as it must have assumed if once in a fused condition, and that its present want of density can be accounted for only by the highest conceivable interior temperature.

Zoology teaches that animals began their organizations in low, simple forms, and that these forms have become more and more diversified, as the surface of the earth from time to time grew more diversified by cooling from a melted state.

Botany teaches these same truths respecting the history of vegetable forms, which zoology makes known respecting animals. The early history both of plants and animals indicate a high tropical temperature all around the globe from the influence of the interior heat when the crust of the earth was yet thin.

Chemistry teaches, from the chemical composition of the earth, that in the beginning it must necessarily have been all on fire, first in a gaseous and then in a liquid state.

From this large number of very widely different sources, the conclusion is irresistible that formerly our earth as a star shone with its own independent light, and that chemical action, the combination of its simple chemical elements, was the cause of that light.

But what causes chemical action? This we do not know. We are here left to conjecture, to theory, and to probabilities. As light and heat act through some media, the vibrations of which run with the velocity of twelve million miles a minute; as electric and magnetic attractions are also caused by ethereal media; so we suppose that chemical attractions must be caused by an ethereal medium, whatever it may be. But from our small laboratories, or artificial furnaces, we can form but a faint and very partial idea of what chemical action must be on so large a scale as our globe, 25,000 miles in circumference, or as the sun, nearly 3,000,000 miles in circumference. When we behold a thunder-cloud, alive with electricity, every instant glowing with flash after flash, we can form but a very imperfect idea of what is going on in that cloud from our experiments on our small electric machines. In like manner, when we have evidence that the primitive heat of our globe was caused by chemical action, we must have only a very imperfect idea of what chemical action really is on such a grand scale. Therefore, in saying that the former heat and light of our globe was caused by chemical action, we must regard that action as in a very extended sense, and modified differently from any thing we have ever seen or thought of. Not knowing the cause of chemical action even in our little laboratories, we must necessarily assume that on so grand a scale as in a burning star, that cause may show itself with great differences of action, especially with respect to the intensity and to the prolongation or continuance of that action. This continuance or prolonged duration of chemical action in a star, may depend on several possibilities, which I shall point out in another chapter.

In quite recent times the opinion has been generally held that chemical action in a star, as the cause of its light and heat, must necessarily be very brief—too brief, for instance, to be the cause of the light and heat of the sun. This opinion I will endeavor to show is rash and founded on several unwarrantable assumptions. It assumes that the unknown cause of chemical action can operate with no greater intensity and no longer duration in an immense laboratory like the sun than in our artificial combinations. It assumes that the materials of the sun and fixed stars have no greater light and heat giving power than those in our earth, even when it is plain that those in our earth vary widely in their heat and light giving powers. It assumes that, in the great laboratory of the sun, the process of creating new chemical elements, new modifications of matter out of older forms, has ceased. All these assumptions I will try to show are without proof and unscientific, and therefore the opinion founded on them is worthless. The only sound and profitable course is to collect and examine all the facts within our reach respecting the light and heat of the sun and of the other stars, whether they be at the present era luminous or unluminous. We have laid a firm foundation for our superstructure in the facts just collected respecting the heat and the former light of the earth. It was due that we should begin with the star on which we live. Its chemical action has ceased, and it is left unluminous. About this there can be no doubt. Our next procedure would be to examine our next nearest neighbor, the moon. But there the fires have long since been extinct; and even its interior heat has so far radiated away that the volcanoes there, once numerous and grand, are now all inactive and dead. Being smaller, and without a thick, non-conducting atmosphere, the moon has progressed in cooling a long stage beyond our earth. Nevertheless, as we will see in a future chapter, the points of resemblance between the moon and the earth show that they have passed through the same fiery process. The inequalities on the moon's surface, the deep depressions in some regions and the elevations in others, correspond to the high table-lands and the low ocean floors of the earth, and in both stars these elevations and depressions are referrible to the same igneous causes. The many volcanic craters, the long lines of lava currents, and the steep mountain ridges, tell on the moon's former igneous history the same as they do on that of our globe. The causes of their fused and luminous condition must have been the same. It is most unphilosophical to suppose that different stars were lighted up by totally different causes. They are numbered by hundreds of millions, their light is of the same nature and subject to the same laws, and the causes of their light have doubtless been the same in each one; whether that light has long been extinct as in our earth, or whether we see it become extinct in modern times as in lost stars, or whether as in the sun it still glows as lively as ever.

The next object of near neighborhood and of investigation is the sun. There we behold a seat of the most unexampled activity. In that extraordinary activity there are many distinct and well-defined facts. These we will bring together and examine, one by one, and the wonder will strike us strongly that they all belong unmistakably to chemical action.

After the sun we will enter on the investigation of the fixed stars. There also we will find many facts, well and clearly ascertained, which are easily explainable by chemical action, and by that alone.

After these examinations of facts we will attend to the philosophy of the subject, at least so far as to show that chemical action must not necessarily in all cases be as short lived as in our laboratories. I hope then to make evident that in different materials there are almost infinite differences of light and heat giving power; and that the materials of the sun and fixed stars being different from ours, their heat-giving powers may last infinitely longer. I hope to show that the unknown cause of chemical action, very probably, or certainly, acts with different modifications, when on so extended a scale as the sun, a laboratory 882,000 miles in diameter; and that with these different modifications the amount of heat and light may be very greatly intensified and prolonged. I hope also to show that what we call the simple chemical elements are now actually in the process of formation in all great contracting globes, which according to the nebular theory are still in their process of condensation; and hence with the formation of these fresh elements there are daily supplies of new fuel. Even though I may not show all these things with absolute certainty, still if I can show merely their probability or their possibility, then the objection against chemical action as the cause of the heat and light of the stars must fail. That objection is founded on the assumption that chemical action must necessarily be as short-lived in those immense condensing globes as in our own furnaces. If it can be shown that this is not necessarily true, but that there *may be* a provision for a longer duration of chemical activity, then the objection vanishes, and we must rely on the many positive evidences in the earth, in the sun, and in the fixed stars, that chemical action has been really the cause of their light and heat.

SECTION VII.

THE CAUSE OF THE LIGHT AND HEAT OF THE SUN.

BEFORE giving the evidences that the light and heat of the sun spring from chemical action, a few words should be said about two other theories. The first of these, and the one more widely diffused in books, regards the sun as a dark body like our earth. On its surface reposes a transparent atmosphere, and high up in this floats a thick, cloud-like envelope, highly reflective, and nearly impervious to light and heat. At a still higher elevation in the transparent atmosphere, floats another cloud-like envelope, which is self-luminous and the source of all the light and heat emanating from the sun. Above these envelopes the transparent atmosphere extends a great distance, and appears in total eclipses as the corona of the sun. These cloud-like envelopes are called the photospheres, and the dark spots on the sun are supposed to be openings in the photospheres, revealing the dark body of the sun. These openings are supposed by some to be caused by "spiracles" in the body of the sun, air-holes in that great orb, which blow the photospheres aside. According to others, these openings, are caused by great whirlwinds in the sun's transparent atmosphere, similar to the cyclone storms in the atmosphere of the earth.

Two considerations led to the formation and extensive adoption of this remarkable theory. The one is, to render the sun inhabitable for intelligent beings like ourselves. The lower envelope is supposed to serve as a screen to ward off the intense heat radiating from the higher envelope, and thus to give a mild temperature to the sun's solid surface. But we know too little, both of final and of physical causes, to propose a special apparatus for rendering the sun's surface inhabitable for beings like men. Our earth was for countless millions of years uninhabited by men, and we cannot say when the proper time should come for peopling the sun. The other consideration, which led to the formation of this theory, was to account for the spots on the sun. But these spots, I shall endeavor to prove, may be accounted for by a much more natural method, and without a resort to any thing so extraordinary or so out of the way of all known facts as these cloud-like photospheres. The objections to this theory are numerous and powerful, some of which may here be stated.

1. We know of no cause for the great amount of light and heat in this photosphere. That amount is immense, and has endured many millions of years, judging by the facts of geology. And we not only know of no cause for this heat, but we can conceive of none. It cannot come from electricity, for we can think of nothing to excite that electricity. The upper photosphere cannot be compared to our aurora borealis, because our northern light is caused by disturbances in our atmosphere arising primarily from the

heat of the sun. It owes its origin therefore to a well-known foreign body. We can conceive of no such origin to a photoshere around the sun. Our aurora borealis is so exceedingly faint as not to be seen in the day time, and therefore it is not comparable to the mighty blaze of dazzling light in the sun. Light and heat cannot exist without a cause. The most mighty of all causes must be necessary to produce the great light and heat in the sun. But here we have a theory for the production of light and heat without any cause. The theory says that the light and heat come from the photosphere; but there is no imaginable way by which they can first get in the photosphere. If we are required to believe that the earth stands on the back of an elephant, we have a right to inquire what the elephant stands on. The truth is, that the photosphere theory is only putting off the difficulty one degree further. Some new auxiliary theory must be gotten up for putting light and heat in the photosphere, and this cannot rationally be done. The Hindoo sages say the elephant stands on a turtle, and the European sages say the light of the photosphere comes from electricity; but then again we must inquire what does the turtle stand on, and *what power* excites the electricity that excites the light and heat? This power cannot even be imagined, much less can it be proved or rendered probable.

2. The causes for these openings in the photospheres, revealing the supposed dark body of the sun, are strange and most unlikely. Why should there be "spiracles" in the supposed dark body of the sun? The only reason ever assigned for their existence is, to blow aside the theoretical photospheres. But this is building a theory on a foundation too airy for belief. The other cause for the openings in the photospheres arises from great whirlwinds, like cyclone storms in our atmosphere. But the sun-spots are more or less irregular, some of them exceedingly irregular; and if

they were caused by a cyclone whirlwind, they would all whirl, and be plainly seen to whirl, without exception. Still, not one of them has ever yet been observed to make a single rotation.

3. The sun-spots break in pieces, and occasionally they have bright cracks and wide, bright openings in or near their centres. When a spot breaks and its fragments separate with great velocity, the advocates of the photospheres floating in a transparent atmosphere, say that narrow bridges of the photospheres have crossed the openings. But the cause which blew aside the photospheres, must surely blow these delicate bridges away. And no such explanation can be applied to the bright cracks and wide, bright openings in the middle of some of the spots. These latter cases are wholly irreconcilable with that theory. But they are easily explainable by the theory of chemical action. Regarding the spots as temporary solids floating on the fiery liquid body of the sun, we can easily understand how by violent agitations these solids may break and fly apart, and how openings may be broken in their centres where the melted and fiery body of the sun may shine through. Afterwards these solids, from causes soon to be given, may dissolve and disappear.

4. By the large number of facts brought together in the beginning of this volume, it appears that our earth is a star of the same nature as the sun and the other fixed stars. We must therefore believe that its former light and heat, and the present heat of its interior, were produced in the same manner as in the sun and fixed stars. But we cannot conceive how our earth could be fused by a hot luminous photosphere floating in the atmosphere. We cannot conceive how such a photosphere could have got there, and where it has since gone. Moreover, according to theory, the lower photosphere of the sun is a dense envelope, im-

pervious to light and heat, and thus rendering the surface of the sun cool and habitable. Therefore, such an apparatus could not account for the present interior heat of the earth, nor for the former fusion of its entire crust.

5. This theory can give no account for the dark, fixed lines in the solar and stellar spectra. These lines are caused, beyond doubt, by metals and other simple chemical elements vaporized in flame. But we can conceive of no way by which the metals could get in that high upper photosphere, floating loftily in the sun's transparent atmosphere.

For these reasons, the theory of luminous envelopes floating in the sun's atmosphere must be rejected.

The other theory for the light and heat of the sun has been quite recently proposed by Mayer, of Germany. He supposes that meteoric stones are falling in the sun, and by their violent collisions they produce the light and heat. This is striking fire, after the old manner with flint and steel, on a grand scale. The objections to this theory are these:

1. The constant accumulation of such materials, during hundreds of millions of years, would increase the body of the sun and its consequent gravity so greatly as to derange the entire solar system, by destroying the balance between the centripetal and centrifugal forces now acting on the planets. Mayer attempts to avoid this objection by supposing that the vibrations of light and heat emanating from the sun, carry off the matter of the sun, and thus counteract the accumulations of meteorites. But this is contrary to the action of the vibratory theory, and agrees with that of the old corpuscular theory, which is now universally given up. The sun need lose no more matter by the vibration of its atoms than the ocean of our earth by the agitation of its waves. If, perchance, some few particles fly off as light and airy spray, gravitation will bring them back again.

2. As we must believe that all stars were lighted up by the same means, so we must believe, according to this theory, that the present interior heat of the earth and its former melted condition in both exterior and interior, was caused by the fall of meteorites. But if so, they must have gradually ceased to fall, as space became cleared of their presence, and we would now find a thick covering of meteorites on the earth's cooled surface. Instead of this, we find them very rarely, and in accordance with their present very rare falls. Mayer obviates this objection by supposing that two great globes came in collision, and thus produced a fusion and a flowing together of both into one. But the known stability of the solar system disallows any such assumption. Its admirable mechanism provides against any such catastrophe as a collision of planetary bodies.

3. Our moon shows evidences of having been in a former melted condition; but we see no evidences of meteoric accumulations on its surface, as there must have been with the necessary gradual cessation of their fall. The cessation of their fall must have been occasioned by the clearing up of the realms of space around the moon, and this clearing up or exhaustion of the meteorites, must have been gradual.

4. The increased masses both of the earth and the moon, by the fall of so many meteorites, must have increased their gravity, and deranged the balance between centripetal and centrifugal forces. Thus the system of the earth and moon, as well as the general solar system, would have been destroyed.

For these four reasons, the meteoric theory cannot be received.

We now come to the old theory, that the light and heat of the sun arise from a great fire, a vast conflagration. This idea, I believe, has been more or less prevalent from time immemorial, and latterly it has been called the theory of chemical action. It was believed by such great men as Newton and Laplace; and the most eminent of all British chemists, Davy, proposed an extension of this theory to account for the former entire fusion of our earth. Newton, in the third book of his "Principia," in announcing "Rules for reasoning in Philosophy," says in his second rule: "Therefore to the same natural effect we must, as far as possible, assign the same cause, as to respiration in a man and in a beast, the descent of stones in Europe and America, *the light of our culinary fires and of the sun*, and the reflection of light in the earth and in the planets."

Again he says: "Fixed stars that have gradually wasted by the light and vapors emitted from them for a long time, may be recruited by comets that fall upon them, and from *this fresh supply of new fuel*, these old stars, acquiring new splendor, may pass for new stars. Of this kind are such fixed stars as appear on a sudden, and shine with a wonderful brightness at first, and afterwards vanish little by little. Such was that star which appeared in Cassiopeia's chair in 1572."

Laplace, in his "Système du Monde," p. 54, says: "As to those stars which appear almost suddenly with great brightness and then disappear, we may suspect with reason that *great conflagrations*, occasioned by extraordinary causes, have taken place on their surfaces. This suspicion is strengthened by a change in their color, analogous to what occurs upon our own planet in bodies which, having become incandescent, subsequently undergo extinguishment."

Thus both Newton and Laplace, resting on the axiom that the light of the sun and of our ordinary fires, is from the same cause, explained the sudden blazing forth of a temporary star as a great conflagration occasioned by extraordinary causes adding a fresh supply of new fuel.

Davy, being the first to decompose the earths and to prove that they are composed of a metal and oxygen, was the first also to announce the theory that the former fusion of our planet was caused by the combination of its simple chemical elements, which are merely other words to describe a celestial body wrapped in a great conflagration, shining and radiating heat like the sun.

But this theory of chemical action in the sun has long since been universally abandoned on account of the groundless assumption that the sun and other fixed stars have no fuel of the right kind to keep up their burning so long. The originators of this assumption do not say how they got their information. It has doubtless come by the most distant and flighty analogical reasoning, and once having gained firm possession of the general belief, it will hold on tenaciously like all inveterate general beliefs. It can be removed, however, in two ways: first, by showing that it is a mere assumption without proof; and secondly, by advancing the strong evidences in favor of chemical action now going on in the sun.

These strong evidences are the following: The sun consists of the simple chemical elements analogous to those in our earth, and therefore it is made up of the most combustible materials. It is surrounded by a coating of flames from two to four thousand miles high. These flames, like all others, are in constant agitation, and here and there send up flakes much taller than the rest. As a supporter of combustion the sun is surrounded by an atmosphere visible to us nearly half a million of miles high, and therefore its entire probable height, including its invisible portion, must be two millions of miles. In this atmosphere float immense clouds, some are eighty thousand miles high, and some are eight hundred thousand miles in breadth, which clouds may be regarded as the smoke and vapor of the great conflagration.

The body of the sun below the flames is a melted liquid mass, in which liquid the chemical action or burning is chiefly taking place, and which, like our ocean, is disturbed by general and special currents. On this great shoreless ocean float here and there dark-looking spots, which are solidified portions, and which soon melt again by the intensity of the heat. This heat and this light of the sun are appropriate products of so vast a burning. Such an array of proofs is very strong, not only from the individual weight of each one, but from their assemblage in one object, their blending together in one locality. To the establishment of these proofs we will now proceed:

1. The fixed lines in the solar spectrum prove the existence in the sun of those modifications of matter which, in the science of chemistry, are called simple elements. The union of these and of their compounds, constitutes chemical action, one of the forms of which is combustion or burning. The fixed lines in the sun's rays number many thousands, in fact they are innumerable; this indicates that there must be many thousand simple elements in the sun. Hence the *materials* for a great conflagration are far more abundant in the sun than in the earth.

2. It has been proved that the visible luminous surface of the sun consists of flame; that is, of vapors or gases so highly heated as to give out light. Incandescent solids or liquids, whose surfaces are seen at an acute angle, give out polarized light. The sides of the sun are seen by us at all angles, but no polarized light is perceived, and hence it appears that incandescent gases or flames envelop the great orb of day. The same fact is proved by a very different method. The dark fixed lines in a spectrum of light can proceed only from a flame, hence we have a double evidence of an envelope of flames around the sun.

The height of these flames, from two thousand to four

thousand miles, must be equal to the depressions of the dark spots which are down in the surface of the sun, and these depressions have been ascertained in the following way: The spots are never seen on the extreme border of the sun. If they were on the exact surface, then, as the sun rotates, they would be seen first on the extreme eastern, and last on the extreme western border. But they do not; they are hidden by the tall flames both on the east and on the west, and we can see them down between the flames only when the side of the sun on which they are placed is turned toward us. One occasion only is reported when the spot was so large that it reached the extreme border, and then formed a notch or indentation on that border. The spot was reported as of an extraordinary size. The indentation may be understood as a gap, where the flames for a time were obliterated by the spot. This, however, is but an extreme case of the kindred fact that the larger the spot the more nearly it can be seen toward the sun's border; and the reason of this must be found in the depression of the spots beneath the luminous envelope. Again, when a spot comes into view on the eastern border of the sun, the first part seen is the eastern side of the penumbra, then appears the nucleus or dark central part, and lastly the western side of the penumbra. The reason is, because the side of the spot toward us is hidden behind the uprising wall-like mass of flame. As the sun turns round, we see the penumbra all around with the central nucleus, because we look straight down parallel with the walls of the flame. As the sun rotates still further, and the spot advances toward the western border, we are unable to see the eastern side of the penumbra because it is nearest to us, and in its turn hidden behind the tall flames. The nucleus next disappears, and lastly the western edge of the penumbra. Dr. Wilson, professor of astronomy in the University of Glasgow, was the first who announced, in 1783, that the spots were below the general level of the sun's surface; and after spending some time in the investigation, he concluded that their depression in several instances was from two thousand to four thousand miles. I do not know that any other estimation has been made, which is very much to be regretted. The subject is worthy in the highest degree of renewed investigation, with improved modern instruments.*

3. The body of the sun, immediately below the flames, consists of an incandescent liquid. When the fixed lines from the vapors of any elements, are seen by the light of the flames in which they are suspended, then these lines are bright; when seen by a more powerful light shining through those flames, then the fixed lines are dark. In the solar spectrum they are dark; therefore they are seen by a stronger light coming from beneath the flames, and hence from either a solid or a liquid. It is proved to be a liquid by its intense heat and by its currents, as will be shown when we come to the spots.

4. We behold great agitations among the flames of the sun, the same as in all great conflagrations. The surface of the sun is the most unquiet of all known places, as might be supposed in a chemical action among so many and such large amounts of elements, combustion on so vast a scale. Sir John Herschel, in describing the sun, speaks of that "excessively violent agitation which seems compatible only with the atmospheric or gaseous state of matter."

5. The atmosphere of the earth has an agency in all great burnings, both in supporting combustion, and in receiving and removing the extinguishing gases. The sun also has an enormous atmosphere, extending upward from his surface probably a million or two of miles. It is seen only in total eclipses, and then called the corona. In the

* See last Appendix.

eclipse of 1851 it was computed by Professor Airy to be little less than the moon's diameter. A breadth of the moon's diameter at the distance of the sun would be nearly a million of miles. This, however, was only the visible portion of the sun's atmosphere, that region nearest the sun, and loaded with vapor. The visible portion of the earth's atmosphere is less than one-tenth of the whole. Other observers have estimated the breadth of the corona to be only one-half, and still another observer thought it to be only one-third of the diameter of the moon. From all the observations in the four eclipses in which the corona has been estimated, we may take more than half the moon's diameter as a low mean for the breadth of the corona. Then, assuming that only one-fourth of the sun's atmosphere is visible, the entire height would be about two million miles. The moon has no appreciable atmosphere; that of the earth, according to ordinary estimates, extends upward fifty or sixty miles; according to more recent estimates derived from meteors, its height is about two hundred miles. The atmospheres of Venus, Jupiter, and Saturn are so high and dense that separate spots on the bodies of these planets can scarcely be seen, and hence the periods of their rotations have been made out with difficulty. From the moon to the sun, as the two extremes, there are great varieties of atmospheric heights, and our only safe guide in forming our opinions in any particular case is direct observation, connected with the fact that atmospheres are visible only in part, and this chiefly through cloudy vapors in their lower strata.

6. From great conflagrations we must expect great clouds; and the clouds in the sun are on a scale corresponding to flames from 2,000 to 4,000 miles high, and to an atmosphere such as I have just described. One cloud has been represented by a most reliable observer as extending perpendicularly 3', and another as reaching in breadth

nearly one-third around the sun. This would make the height of some clouds 80,000 miles, and the breadth of others 800,000. Probably still larger clouds occasionally arise; for observers have seen them for a few moments only on four occasions, in 1842–'51–'58–'60, and it is not likely that in these transient glimpses they have happened to meet the highest or the broadest. Such vast dimensions should not seem extravagant. On the contrary, they are entirely in keeping with the circumstances. These tall flames, the atmosphere, the clouds of smoke, the combustion on so grand a scale, producing such an amount of light and heat, all perfectly harmonize.

On the earth some gases and vapors are almost infinitely more transparent than others, and it may be that those forming the clouds in the sun are naturally not very opaque. The intense heat of the sun may render them more transparent, the same as heat renders the vapor of water perfectly transparent. We therefore perceive the reason why the powerful light of the sun pierces so completely through those clouds that we are unable to see them except in total eclipses. These clouds cannot form the spots in the sun, for they are so much larger than the spots; and the clouds are elevations on the sun, the spots are depressions.

The probable origin of these clouds is obvious. The products of flames are always gaseous, either nearly permanent as carbonic acid, or easily condensible as the vapor of water. The products of the deep flame-coating around the sun must be inconceivably abundant. Their partial condensation must form vast clouds. If these clouds did not further condense and form precipitations, then they must increase continually and at length shut out the light of the sun. But we have no such evidence of their increase, and hence we are brought to the conclusion that the condensible gaseous products of the flames must fall in some form back again down through the same flames. The forms of the precipitations of the single vapor around our globe are so various—as dew, fog, mist, rain, hail, snow, frost—that we cannot begin to decide on the forms of precipitations in the sun. When iron is burned in a glass jar containing oxygen, the product very soon assumes the form of a fine impalpable powder, which slowly settles on the sides and the bottom of the jar. But this small, dry affair teaches us very little of the form of the metallic compounds as they fall back into the body of the sun. Imagination, not science, may here have its full play among metallic rain, and hail, and snow. Fancy, not philosophy, is at liberty to picture the metallic hail falling from the tops of clouds 80,000 miles high, and hastened down, not by a feeble attraction like that of the earth, but by the great gravitating force of the sun.

7. In all great conflagrations the flames are grouped together in masses here and there, more bright and voluminous than the rest. They are never smoothly and evenly distributed. The sun has its brighter and more intensely lighted portions, special masses and ridges of flames; and they are constantly changing. "In the neighborhood of great spots or extensive groups of them, large spaces of surface are often observed to be covered with strongly-marked, curved, and branching streaks more luminous than the rest, called faculæ, and among these, if not already existing, spots frequently break out. They may, perhaps, be regarded, with most probability, as ridges of immense waves in the luminous regions of the sun's atmosphere, indicative of violent agitations in their neighborhood." *

The proof that these brighter lines are really ridges of flames has been furnished by actual observation on at least two or three different occasions. The Rev. W. R. Dawes, of England, with a refractor of 8½ inches, and a power of

* Herschel's "Outlines."

234 diametres, had a fair view of one of these bright ridges running parallel to the border of the sun's disk, and when it precisely occupied the border. He "satisfied himself that it *projected irregularly beyond the circular contour* formed by the edge of the sun. To obtain ocular demonstration of the bright streaks being really elevated ridges or waves in the exterior luminous envelope, is, of course, a very rare occurrence; but, in the present case, the evidence was as complete as could be desired. The combination of circumstances was most fortunate; a bright streak of unusual size being precisely at the sun's edge, and the state of the air permitting the use of a large aperture and a high power with full advantage." *

The other occasion was not when the streak was parallel with the edge of the sun, but when two streaks, one each side of a large spot, ran at right angles across the edge of the sun's disk, and formed two elevated prominences on that edge, leaving a gap between them. †

8. So great an orb as the sun, nearly three million miles in circumference, and burning with flames a few thousand miles high, should send forth an inconceivable amount of heat, and accordingly we find that the actual heat of the sun is on a corresponding scale. Its intensity is proved by its penetrating power, easily passing through glass and other substances which greatly obstruct our low degrees of artificial heat. Still the real amount of heat given out by the sun has never been determined. Different estimates have been made, from certain experiments, but I have never been able to see their conclusiveness. We can say, however, that guided by our common sense, and so far as our knowledge goes, we can see no disproportion be-

---

* "Monthly Notices of the Royal Astronomical Society," vol. xx., p. 56.

† "Monthly Notices," vol. xxiii., p. 18, also p. 110.

tween the heat we receive from the sun and a conflagration so grand. The two accord perfectly together.

9. The intense flood of light from the sun is another great fact corresponding to all the others we have already enumerated. The light, the heat, the flames, the agitations and high ridges in these flames, the inflammable chemical elements, the melted body of the sun, the clouds of smoke, the vast atmosphere, are the best possible evidences of a great conflagration. What else would we have? What else can they be? We know of nothing else that they can be, and I do not see how we can think of stronger proofs of a great burning. They would be at once admitted, were it not for a single objection. The objection is, that there is too much heat for the amount of burning materials in the sun. It goes upon the assumption that no materials could possibly be created to give out more heat than such as we have here in our little planet! This assumption is most extraordinary. It is as much as to say that our globe, 8,000 miles in diameter, circulating around the sun between Venus and Mars, the goddess of Love and the god of War, is in all respects the *ne plus ultra!* We thus make it the standard by which to gauge even the chemical elements of the sun, and say, So far shalt thou go and no farther. And we do all this in full view of the different relations to heat of the different elements even in our own globe. One gives out little heat, and another much; one absorbs a small amount of heat, and another a large amount. A pound of water requires thirty times more heat to warm it than a pound of mercury. In all other respects there are wide differences between the elements here, in their densities, their combining numbers, their electric, their magnetic, and all their other relations. The planet Mercury is at least sixty times more dense than one of the satellites of Jupiter. If we make the very easy admission that there are probably

wider differences between the materials of the sun and those of our globe than there are between the several substances here, then all the mystery is ended; the objection vanishes. If we be convinced that the materials of our earth have not a heat-giving power sufficient to supply the vast outgivings from the sun, then, in view of all the evidences of a great conflagration in the sun, it would be more philosophical to conclude that the materials of the sun, pound for pound, produce more heat than our own. There are at least two other ways of proving the unsoundness of the one objection against the chemical theory; these we will attend to shortly. We now go on with the remaining facts from the sun in favor of that theory.

10. The general surface of the sun between the bright ridges of flames, is thickly strewed over with small darklooking specks in a state of change, appearing and disappearing. They seem like mere points at this distance, and convey the idea of chemical action. The part of the sun's disk not occupied by the spots, and the bright ridges, "is far from being uniformly bright. Its ground is finely mottled with an appearance of minute dark spots or *pores*, which, when attentively watched, are found to be in a constant state of change. There is nothing which represents so faithfully this appearance as the slow subsidence of some flocculent chemical precipitates in a transparent fluid, when viewed perpendicularly from above." * Schwabe, a very reliable German observer, gives the following description: "The fainter portions lying between the vein-like luminous clouds—'elevations of flame'—on the general surface of the sun, are deeper depressions and always present a shagreen-like gray, and sand-like appearance, reminding the observer of a mass of uniformly-sized grains of sand. On this shagreen-like surface we may occasionally notice exceed-

* Herschel's "Outlines."

ingly small faint gray, not black, *pores*, which are further intersected by very delicate dark veins. *These pores, when present in large masses, form gray nebulous groups, constituting the penumbra of the sun-spots.*" Here we learn the origin of the large spots.

11. These "sun-spots," dark-looking patches on the sun's disk, and surrounded by a less dark border called the penumbra, appear very commonly in the near vicinity of the brighter ridges of flame. Thomas Dick says: "For several years past, when any of these faculæ or ridges have appeared on the eastern margin, I have uniformly been enabled to predict the appearance of a large spot or two within the course of twenty-four or thirty hours, and in more than twenty or thirty instances I have never been disappointed." These spots are probably light, but seem dark by reason of the greater brightness of the sun's general surface; just as the brightest artificial lights, when held up before the sun, seem dark. When compared with the planet Mercury, in its transit over the sun, they seem the brighter. Dr. Henry, since appointed Secretary of the Smithsonian Institution, found, by projecting the image of the sun on a screen, that the spots radiated less heat than the brighter portions of the sun. The sizes of the spots vary from the smallest perceptible pore which, to be visible, must be from four hundred to five hundred miles in diameter, to others which have been 45,000 miles in diameter. Generally they are roundish, with some irregularities, sometimes angular, and occasionally of irregular fantastic shapes.

The sun is seldom entirely free from spots, and in some years they are especially numerous. Dick says: "On the 16th of November, 1835, I perceived about ten different clusters, and within the limits of two of the clusters sixty different spots were counted, and in the whole of the other clusters about sixty more, making in all about one hundred and twenty spots great and small. On other occasions I have counted one hundred and thirty and one hundred and fifty spots. Such a number of spots are generally arranged in ten or twelve different clusters, each cluster having one or two large spots, surrounded with a number of smaller ones. The duration of the spots is various, from twelve hours to six or seven weeks."

The changes which the spots undergo, and their history from their first appearance until they vanish, throw much light on their nature and on the constitution of the sun. As already said, the pores, when collected in masses, form the penumbra, and this is the beginning of the spot. As it grows and enlarges, the dark nucleus begins to appear in the centre. This nucleus increases in size as the penumbra increases, and when the penumbra contracts, the nucleus contracts simultaneously. "During the process of diminution," says Dick, " the penumbra encroaches gradually on the nucleus, and it sometimes happens that the encroachment of the penumbra divides the nucleus into two or more parts. When the spots disappear, the penumbra continues for a short time visible after the nucleus has vanished." " In cases of disappearance the central dark spot always contracts to a point, and vanishes before the border." * It is therefore evident that the penumbra is the most important element in the spot. It has an independent existence; for it is seen before the nucleus appears, and after the nucleus has faded away. The nucleus would seem nothing more than the penumbra, whatever it be, turned dark. Sometimes a penumbra has not only one or two, but even six or eight dark patches toward its middle. The color of the penumbra, though lighter than the nucleus, is not always uniform, some regions being darker and others lighter even in the same spot.

* Herschel's " Outlines."

A very important fact in their history is that sometimes, though rather rarely, slits or cracks appear in the dark nucleus, and bright lines of light shine through. In the figures given by observers we see the precise appearance of shrinkage cracks in drying mud, only they are not so numerous; they seem as if the nucleus had contracted within the penumbra, and subjected itself to these fissures. A picture of a large spot is given in the "Monthly Notices of the Royal Astronomical Society," vol. xxiii., p. 110, where, instead of a narrow crack in the interior of the nucleus, there appears a large bright patch. It is as if one of the cracks had been violently widened, and its edges broken away by the upward force of the melted boiling material below, which is showing its splendor through the broad opening made in the spot.

Another very decisive fact in their history is their sudden cracking and breaking into fragments like a cake of ice on an agitated water. "Occasionally they break up, or divide into two or more, and in these cases offer every evidence of that extreme mobility which belongs only to the fluid state." * Dr. Long, in his "Astronomy," vol. ii., states that "while he was viewing the image of the sun cast through a telescope upon a white paper, he saw one roundish spot, by estimation not less in diameter than our earth, break into two, which immediately receded from one another with prodigious velocity." The Rev. Dr. Wollaston, when viewing the sun with a reflecting telescope, perceived a similar phenomenon. A spot burst in pieces while he was observing it, "like a piece of ice, which, thrown upon a frozen pond, breaks into pieces and slides in various directions."

When a spot is growing, all its lines, those between the nucleus and the penumbra, and outside bounding the pe-

* Herschel's "Outlines."

numbra, are clearer, more distinct and continuous when the spot begins to diminish, then the lines become fainter, more hazy, broken, and jagged.

Immediately around the spots the surface of the sun is brighter and more luminous than its general face, giving the spots, especially when large, the appearance of a bright border.

All these well-observed and well-authenticated facts respecting the spots, seem to me to furnish ample evidence of their nature, especially as all the facts relating to the spots, all the facts in the physical constitution of the sun and fixed stars, all the facts in the formation of the earth and the solar system, point with most singular harmony to one and the same conclusion. The conclusion is this:

The spots are solid bodies floating in the midst of the flames on the liquid surface of the sun, and they are formed of incombustible chemical compounds. They are solid, because they are liable to crack and break. No gaseous, no liquid bodies, can possibly present such appearances of straight and angular fissures—sharply defined, with the clear light shining through. If they were not solid, they could not present the appearances of cracking and breaking into many pieces, and these pieces flying asunder like ice on agitated waters. They are incombustible, because they evidently do not burn in the midst of all that heat. In their beginning, when very small, they are the aggregations, according to Schwabe, of minute pores—minute to us, but to be seen at all they must be a few hundred miles in diameter. If these pores and other small spots were combustible, they would not grow to a large size, they would be consumed. When they finally disappear, it is, as I shall try to show, by melting. These incombustible solids are chemical compounds; if they were not, if they were uncombined simple elements, they would burn, for universally the nature of these is to unite and thus produce burning But after they are united they are incombustible, like stone or water, until by some special process they are again separated. The ground beneath our feet, the general crust of the earth everywhere, is in this chemically compound incombustible condition, and so great a globe could not possibly have got in this condition but by burning.

The dark nucleus in the centre of the spot is merely the penumbra losing its brightness, because it is slightly cooling, and cooling because that region was formed first, and has had the longest time to radiate away its heat; it must also be the thickest, and therefore less affected by the heat below, and it is farthest removed from the surrounding flames. These spots, as already stated, have been proved to be less hot than the flames. We can thus understand why, as the penumbra increases in size, the nucleus increases, and why, as the penumbra begins to melt and decrease, the nucleus also resumes its lighter color, and disappears. We can also understand why the penumbra, as it encroaches on the nucleus, occasionally divides it in two, and also why in the same spot there may be several nuclei. The spots may not in all cases be everywhere even in thickness. The central parts, which are the thickest, must be the darkest, for their external surfaces on account of this thickness are best protected against the interior heat of the sun. If there be two or more portions thicker than the average, there may be on that account two or more nuclei. For a similar reason we can perceive why the penumbra in different portions has lighter and darker shades of color, the thicker portions having the darker shades.

We can readily understand the effect upon the sun when the spots grow to be large. A spot equal in size to the continents of North or South America, or to those of Africa and Asia, with Europe united, would not be excessive.

THE NATURE OF THE SUN SPOTS. 95

Some, as already said, have been known to be twenty, or thirty, or forty thousand miles in diameter. Now, as these spots radiate comparatively little heat, and as the light and consequently the heat come chiefly from the liquid body of the sun, these spots must obstruct the heat to an inconceivable amount. In proportion to their great breadth, and to their want of radiation, must be the magnitude of the obstruction they oppose. Then the heat must escape in more than ordinary abundance all around the vicinity of the spot. Just here is where the tall ridges of flame arise, and here we see the bright border all around the spot. These bright borders, and these elevated sheets of flame, can thus, from the extra escape of heat, be most naturally accounted for.

Another effect of this obstruction, and consequently this confinement and accumulation of heat below the immense spots, is, according to all known laws of matter, expansion of the liquid materials, upward currents, and great agitations. We must in the sun look for more violent upheavings from beneath the spots than all the earthquake power beneath the crust of our own little planet. No wonder that in the exertions of this mighty force of heat an entire continental spot should be split and shaken in pieces! No wonder that from these upward-boiling currents an observer now and then sees the fragments of a spot fly asunder in various directions with "prodigious velocity."

Another effect of this obstruction and vast accumulation of heat, is ultimately to melt the spots and cause their extinction. Their peculiar appearance as they are about to vanish, or rather the difference between their appearances when they are increasing and when they are decreasing, may be clearly accounted for by our present views. As already said, when a spot is growing, all its lines, those between the nucleus and the penumbra, and those outside bounding the penumbra, are clearer, more direct and continuous; when the spot begins to diminish the lines become fainter, more hazy and broken. The descriptions of observers remind us of the difference between a forming cloud with its outlines full and clear, and a vanishing cloud with its outlines obscure and jagged. The parallel between the spots and the clouds in this respect is striking, and it arises from similar causes in both cases. The beginning, both of the cloud and of the spot, is one of condensation or aggregation; the end is a process of dissolving, melting. We have all seen the torn and obscure appearance of ice in the spring, as it melts on a pond. Often it can hardly be told which is ice and which is water. As it departs or melts away, it is always more or less irregular and ill defined in its outlines, the same as the dissolving cloud or the vanishing spot in the sun.

From the facts before us it appears that the temperature of the sun is not high enough to prevent the solidification of small masses on its surface, but yet high enough to prevent those masses from growing so large as to hinder the escape of an essential portion of heat; and whenever the heat is considerably obstructed and imprisoned, it soon acquires a temperature sufficient to dissolve the imprisoning wall. There is thus a constant struggle between the tendency to solidify and the tendency to dissolve, each one by turns gaining the mastery. Every spot, therefore, exhibits a *cycle* of the hardening and melting processes, beginning at a small point, growing gradually to a maximum, and then gradually waning to a small point again. In the same way there is a cycle of about ten or eleven years, during which a minimum of few or no spots, slowly increases to a maximum of many spots, and then slowly diminishes to few or no spots again. Here also we probably see the alternations in the contest between opposing forces, probably involving chemical operations in the cycle, but altogether unknown as yet. The cycles of the many individual spots appearing during the long cycle of ten years, remind us of small ripples on the backs of large waves. From the intense heat in the sun it is probable that the spots are not dark or unluminous, but really incandescent; hence they appear lighter than a planet in its transit, and hence their indications of heat, though less than in the brighter flames.

12. *Currents in the Liquid Body of the Sun.* These currents are indicated by the spots. It was my intention to introduce here the results of all the published observations and calculations respecting these movements, but as yet such an attempt would be premature. Many more observations, continued during several years, and made on the same series of consecutive days on different sides of the globe, are necessary for a complete understanding of this grand and difficult subject. It is from our ignorance of these currents that there is still an uncertainty about the time of the rotation of the sun on its axis to the amount of eight hours, an uncertainty of several degrees about the direction toward which the axis of the sun inclines, and an uncertainty of half a degree about the amount of that inclination. At present we simply know that there are local and temporary currents, and very probably general and permanent currents. About these latter we are especially in ignorance.* The proofs that there are very violent local and temporary currents are as follows:

First. The breaking of the spots, and the flying asunder of their fragments. These motions are described as of "prodigious velocity," and like the pieces of a cake of ice when thrown on a frozen pond. An adequate and natural cause for such events immediately presents itself to the mind. As the heat of the sun is certainly known to be obstructed by the

---

* When this was written Mr. Carrington's work on the Sun had not been published.

spots, and as the spots are larger than the great spreading continents of our globe, they must cause an inconceivable accumulation of heat below them, and that heat must expand all the liquid beneath the spots, and hence give these liquids a mighty upward impulse. Then it is natural that the solid continental spot immediately over this tremendous boiling, should occasionally be rent in pieces, and that these pieces should be hurled apart by the outward spreading of the vertical current. This vertical current may be hastened by the dilatation, not only from the heat directly, but also from extraordinary chemical processes, occasioned by the unwonted heat. Even without the agency of the spots, and without the evidences they supply of vertical currents, there are well-known natural principles that cause vertical currents in all liquids which radiate much heat from their surfaces, and therefore such vertical currents must circulate in the sun.

Secondly. The approach to each other, or the departure from each other of two spots. In the "Monthly Notices of the Royal Astronomical Society," lately published, an account and drawing are given of a large spot that broke in two, and of the increasing distances apart day by day of the two fragments. In the "Comptes Rendus de l'Academie des Sciences," 1842, tom. xv., p. 941, M. Laugier reports two spots, whose angular distance from each other was 44° 29' on June 29th, 1838, and again on July 2d, their angular distance was 46° 2'. He reports two others, May 24th, 1840, at the angular distance of 78° 30', and on the 27th their angular distance was only 73° 32'. He concluded that their approach toward each other was at the rate of one hundred and eleven metres per second, or two hundred and fifty miles per hour.

Thirdly. Another method of proving the existence of local and temporary currents is by the time which each spot, in or near the same latitudes, requires to make one revolution around the sun. In the Memoir just cited, M. Laugier found the time of the revolution of the sun on its axis to be by one spot 25.34 days, by another 24.28 days, and by another 26.23 days. If we take the first or mean, twenty-five and one-third days, as the true time of the sun's rotation, then the second or shortest would indicate a current of about two hundred miles per hour, in the direction of the sun's rotation; and then also the longest would indicate a current of about one hundred and fifty miles per hour, in a direction contrary to the sun's rotation. All these numbers are very high, from one hundred and fifty to two hundred and fifty miles per hour; they are obtained by two very different methods—one by the approach together of two spots, and the other by the passage of spots around the sun. The harmony between the two methods is remarkable; and yet high as the velocity of these currents appears, it does not convey to the mind as high a velocity as the description in the first method, about the breaking of a spot and the flying asunder of its fragments.

Such very violent, local, and temporary currents must be attended with rough agitations on the liquid surface of the sun. Many other facts indicate the same condition of agitation, as the escape of so much heat and light, the very tall, raging flames, and the powerful action of chemical agencies. The forms of the spots and groups of spots are subject to great changes, and these changes may be explained by agitations in the rough ocean on which they float. A good illustration may be seen in the different features of the same spots, drawn in the "Monthly Notices," vol. xxii., p. 299. In one day several round disks were seen in a group of spots, and on a succeeding day these disks had all preserved their perfectly circular form, except that a single large segment had been cut straight off from their sides, precisely the same as waves in a river or bay cut away great segments from a body of ice.

The constitution of the sun is thus becoming clear and definite to our view. The darkness of the fixed lines proves that the chief source of light and heat is in the body of the sun, beneath the agitated flames. This body is proved to be liquid by its currents, as indicated in the movements of the spots; and the vast amount of heat arising from the sun, proves that the temperature of the sun is far above the point of fusion of all known substances. The spots are only temporary solids, as shown by their cracking, breaking, and ultimate melting, and by other peculiar appearances. The bright external face of the sun is shown to be flame, by its want of polarized light, and by the darkness of the fixed lines. The height of that coating of flame must be a few thousand miles, as is proved by the depression of the spots; such depression is proved by the invisibility of the spots near the apparent margin of the sun, and by the succession of visibility of the opposite sides of the penumbra and of the interior nucleus. The brighter portions called *faculæ* are shown to be tall ridges of flames by ridges and notches actually seen in the sun's limb; and their positions chiefly around the spots, are accounted for by the obstruction which the spots oppose to the heat. The spots are known to be deficient in heat, and hence the confinement of the heat below them, and its escape around their borders, giving rise not only to the extra tall flames, but probably to special chemical reactions, and at last to the dissolution of the spots. The pores, from their great numbers and from their aggregations forming penumbra, aid these evidences of the solidity of the spots. The vast atmosphere of the sun, the lower and visible portions of which rise half a million of miles, must extend in its entire height one or two millions of miles. The clouds floating in that atmosphere, rise occasionally to the altitude of 80,000 miles, and spread out sometimes with a breadth of a few hundred thousand miles. This atmosphere is a fitting supporter of such flames, by carrying off the incombustible and extinguishing gases, and the clouds are becoming appendages of so great a conflagration. The liquid constitution of the body of the sun is adapted to allow the combination or burning of its chemical elements. The extreme agitation both in the flames and in the liquid beneath, complete the evidences of an immense conflagration, affording the light and heat which shine on our earth, and which are the prime causes of all motion here, organic or inorganic, except the tides, earthquakes, and volcanoes.

In comparing the proofs of chemical action now going on in the sun with the proofs reviewed in another section, of a like action formerly in our earth, two ideas impress us strongly:

First. These two classes of proofs are entirely independent, and they are very different from each other; having nothing in common except the fact that the combustible chemical elements abound both in the earth and in the sun.

Second. These two classes of proofs, being independent and different, and leading to the same result, powerfully corroborate each other. From what we now behold in the sun, we are more completely convinced that chemical action formerly put our earth all aflame; and, from beholding the combination of the chemical elements here in our semi-fused planet, we are more completely convinced of the present process of chemical combination in the sun.

There is a third class of proofs in the fixed stars, also independent and different, which equally corroborates these two, and which is also corroborated by them. To that third class we will now proceed.

SECTION VIII.

THE CAUSE OF THE LIGHT AND HEAT OF THE FIXED STARS.

WHILE the proofs in favor of the chemical theory, drawn from the earth, the sun, and the fixed stars, are quite independent and different, there is one proof—the possession of the chemical elements—which is common to them all.

By the fixed lines in the spectra of the stars, we have evidence of the infinite diversity in the stars of those modifications of matter denominated simple chemical elements. These lines differ more or less from each other in every star, and as the number of the stars is infinite, therefore the original matter of creation has been infinitely diversified. This infinite diversification, however wonderful, is really an established principle in the operation of general laws. We see no two faces of men or women, and no two leaves of the forest, precisely alike. The first and most remarkable property of these simple elements, and also of many of their compounds, is to unite and burn. In the fixed stars, therefore, we behold the provision made for chemical action in that form which is called combustion or burning. And the great diversity of these elements in the stars furnishes the fuel not only for so many vast conflagrations, but for flames of every imaginable shade of color; such, in fact, as appear in the stars.

When the simple chemical elements are vaporized by heat, and thus suspended in the flames, as for instance in the flame of a lamp furnace, then the fixed lines of those elements may appear either dark or colored. They appear colored when seen in the light of those flames in which the elements are suspended; and they appear dark when seen in another more powerful light shining from behind through those flames. In the stars these lines are dark, and therefore we conclude with certainty that the fixed stars are surrounded with an envelope of flame so intensely heated as to hold the simple elements suspended within them in a state of vapor, and that beneath those flames a more powerful light comes from the bodies of the stars, which are either solid or liquid. From the movements of the spots, especially as seen in the sun, and from the intensity of the light and heat, we cannot doubt that they are liquid.

Chemical action in the fixed stars is further confirmed by the universal extension through space of a medium or agency producing chemical action. This medium appears to be coextensive with those media producing light, heat, gravitation, and magnetism. When a stone, or a feather, or a drop of water falls to the ground, each one does this by an impulse from the same cause. That cause, whatever it be, we call gravitation. In like manner we conclude that when oxygen combines with hydrogen, or carbon with sulphur, or chlorine with sodium, each combination is effected by the same force. This force is generally called chemical attraction; but we know not precisely what it is; we are certain that it is a medium or agency extending through all space, causing chemical action. It operates by waves or vibrations, like those producing light and heat. The waves come from the sun and from the far-distant fixed stars, and they show their power by producing a picture on a photographic plate. They have been called chemical rays, actinic rays, and actinism. In the sunbeams and starbeams they are associated with light and heat, but they are really distinct from light and heat, because they can be separated by an ordinary glass prism. Thus we behold through all space the universal extension of the chemical medium, and we behold in the stars through all space the existence of the chemical elements. And where this chemical medium and these chemical elements are found together, there we may believe in the possibility of chemical action.

Besides this general argument in favor of chemical action in the fixed stars, there is a large number of special arguments. These will best appear by adopting the following arrangement:

First. The Periodic stars, whose light waxes and wanes periodically.

Secondly. The Irregular stars, the waxing and waning of whose light is not periodic.

Thirdly. The Lost stars, whose light is gone out.

Fourthly. The Temporary stars, whose light has appeared for only short intervals, from a few weeks to a few months or a few years.

Fifthly. Colored Stars, particularly those whose colors have changed. It is of the highest importance, however, to remember that these five classes of stars are made here merely for convenience of discussion. They constitute, in fact, but a single class, for they all graduate into each other by insensible degrees as kindred phenomena. Many of the periodic stars are more or less irregular in their amounts of light and in their periods. Temporary stars are also irregular, and between the variable and the invariable stars the difference is often imperceptible. Both often change their colors.

PERIODIC STARS.

The following table contains twenty-four stars whose periods have been more or less accurately ascertained. The first column contains their names; the second their periods of waxing and waning, in days, hours, minutes, and seconds; the third their maxima, and the fourth their minima of brightness or size:

	D.	H.	M.	Sec.	Maxima	Minima
Algol..............	2	20	48	59	2.3	4
δ Cephi............	5	8	47	39½	4.3	5.4
η Aquilæ...........	7	4	13	53	3.4	5.4
ζ Geminorum.......	10	3	35		4.3	5.4
β Lyræ............	12	21	46	40	3.4	4.5
The Sun...........	25	7?				
β Pegasi...........	40	23			2	2.3
α Hydræ...........	55				2	2.3
α Herculis.........	66	8			3	3.4
Scuti R...........	71	17			6.5 to 5.4	9 to 6
α Cassiopeiæ.......	79	3			2	3.2
Virginis, R........	145	21			7 to 6.7	0
α Orionis..........	196				1	1.2
Leonis R..........	312	18			5	0
Coronæ R.........	323				6	0
Mira Ceti.........	331	20			4 to 2.1	0
Pegasi R..........	350				8	0
Serpentis R........	359				6.7	0
Serpentis S........	380				8 to 7.8	0
Cancrii R.........	380				7	0
Aquarii R.........	388	13			9 to 6.7	0
χ Cygni...........	406	1			6.7 to 4	0
30 Hydra Hevelii...	495				5 to 4	0
ε Aurigæ..........					3.4	4.5
Cancri S..........					7.8	0

The substance of this table is due to Agelander, by whom it was furnished for the third volume of the "Cosmos" of Humboldt.* In two other columns he added the names of the discoverers and the dates of the discoveries of the periods of these stars. He arranged them in the order of the dates of their discovery, but I have disposed them in reference to the lengths of their periods, which throws some light on their nature. Double this number of periodic stars might have been added from other sources; my object,

---

* Agelander, in the "Cosmos," says: "For the purpose of clearly and conveniently designating the smaller variable stars, which for the most part have neither names nor other designations, I have allowed myself to append to them capitals, since the letters of the Greek and the smaller Latin alphabet have, for the most part, been already employed by Bayer."

however, was not to make a complete catalogue of periodic stars, but simply a collection large enough to support my inductions. Many facts, further on, are acknowledged by quotation marks only; they are derived from several other eminent German observers who contributed their manuscripts for the "Cosmos."

The sun I have added to this table, partly because in a telescopic view its spots greatly vary its appearance and make it a variable star, and partly to show the harmony of its period of rotation with the periods of the variable stars. It is slower than some and more rapid than others, the same as we find that its mass is larger than that of some of the fixed stars, as ascertained by the binary systems, and smaller than others, Sirius for example, judging from their amounts of light.

For myself, it is not possible to look on the periods of many of these stars, the first five on the table especially, true and punctual to a minute and even to a second, without the deepest conviction that they are caused by rotation on their axes with one side brighter than the other. There can be no other possible explanation within the present bounds of our knowledge. Even the diversities in the lengths of their periods, some of them less than three days and others more than five hundred, confirm this view; for they are so similar to the diversities of all astronomical phenomena, as seen in the sizes, revolutions, rotations, masses, and densities of the heavenly bodies. While there are such diversities among them in every imaginable way, they are all the same in being true and constant each one to its own period of rotation. This unity in diversity, this invariableness in the periods of well-known rotating spheres, as the sun and the planets, and in the periods of some variable fixed stars, exhibits an identity which is perfect and impressive. The explanation of the invariableness of the one by the invariableness of the other, is simple and complete. After these remarks on the regularity of the periods of some periodic stars, it might seem that this is the proper place to speak of the irregularities in the periods of others. But these irregularities of period and their causes can be better explained after attending to several other classes of irregularities.

IRREGULARITIES OF MAXIMA.—By looking at the table in the column of maxima, two different maxima will be seen for many stars. This indicates that these stars are irregular in their maxima, being sometimes greater and sometimes less. Thus, Mira Ceti occasionally reaches nearly the second magnitude, and at other times it rises no higher than the fourth. This difference may be caused by variations in the sizes or numbers of the dark spots, obscuring occasionally even the brighter side of the star.

IRREGULARITIES OF MINIMA.—The minima of some periodic stars are as irregular as the maxima. In the table many of the minima are marked 0, which indicates that the light becomes fainter than stars of the tenth magnitude. At these very low magnitudes, when the star is no longer visible to the naked eye, and is seen only through powerful glasses, the observations being more difficult are more seldom made. The cause of the difference of minima is, doubtless, the same as that of the maxima, namely, variations in the sizes of the spots.

DIFFERENCES BETWEEN MAXIMA AND MINIMA.—Among different stars the greatest diversities are found; some varying the amount of their light from the second to the tenth magnitudes, and others affording the least perceptible difference between maxima and minima. Scuti R at times diminishes from the fifth down to the ninth magnitude; once it totally disappeared, and occasionally it varies only half a magnitude. Alpha Herculis occasionally " scarcely changes it light for months together." "The variability of the light of Betelguese is scarcely noticeable during some periods, but from 1836 to 1840 they were striking and unequivocal." "Coronæ R is variable only at times:" generally it is not a variable star. About other stars there are disagreements among observers whether they be variable or not. Sir John Herschel classes the temporary star first seen in Cygnus, in the year 1600, among the variable stars; Agelander does not.

PERIODIC AND INVARIABLE STARS THE SAME IN NATURE.—In the above facts we see the easy transition between variable and invariable stars. There can be no doubt but that if we possessed delicate instruments for the measurement of light, our sun would be found to be a variable star; his spots at times are large and numerous, and again they disappear entirely. The same is doubtless true of hundreds of other stars now regarded as invariable. At present our photometric instruments are so defective that our chief reliance is on our imperfect senses. As light and heat are kindred things, we may understand our deficiency in measuring light by the assistance we derive from the thermometer in measuring heat.

IRREGULARITIES IN THE INCREASE AND DIMINUTION OF LIGHT.—When periodic stars increase or decrease the amount of their light, they do this very generally in an irregular way, sometimes rapidly, sometimes slowly, and occasionally standing still for a while. They have thus certain ROUTINES of irregularity, and in SOME STARS THESE ROUTINES OF IRREGULARITY ALWAYS REMAIN THE SAME; IN OTHERS THE ROUTINES ARE OCCASIONALLY ALTERED.

Algol. "The augmentation and diminution of its brightness are not quite regular, but when near the minimum they proceed with greater rapidity. It is, moreover, remarkable that this star, after having increased in light for about an hour, remains for nearly the same period at the same brightness, and then begins more perceptibly to increase."

Delta Cephei increases and decreases with much regularity except in the latter part of its diminution, when, during eight hours it scarcely changes at all, and very inconsiderably for a whole day.

Eta Aquilæ at first increases slowly, then more rapidly, and afterwards again more slowly. When it has reached its maximum its brightness does not diminish quite so regularly, for after it arrives at a certain stage it changes more slowly than either before or afterwards.

Beta Lyræ has two maxima and two minima in one period. The two maxima are equally bright, but one minimum is much fainter than the other. The entire course of its changes can be accounted for by supposing two opposite quadrants equally bright, and between these, other two opposite quadrants unequally dark. Then by a revolution there would be two maxima equal in amount of light, and two minima unequal.

Epsilon Aurigæ shows an "alteration of its light either extremely irregular, or else in a period of several years there are several maxima and several minima; a question which cannot be decided for many years."

In all these instances where the routines of irregular increase and decrease of light remain the same constantly, we may account for the phenomena by supposing that the spots around the star are placed irregularly, thus intercepting the light irregularly, and that they always keep the same sizes and positions. But if their sizes or positions should change, then *the routines would change.* The following is an instance:

Mira Ceti when first visible to the naked eye increases in brightness with great rapidity, afterwards more slowly, and at last with a scarcely perceptible augmentation; then again it diminishes at first slowly, and afterwards more rapidly. Occasionally this star, at the period of its greatest brightness, exhibits for a whole month together scarcely a perceptible variation; at others a difference may be observed in a very few days. On some occasions, after the star has decreased in brightness for several weeks, there has been a period of perfect cessation, or at least a scarcely perceptible diminution of light during several days; this was the case in 1678 and in 1847. The above explanation easily accounts for these *changes of routine*, namely, changes in the spots.

INEQUALITIES IN THE PERIODS OF THE INCREASE AND DECREASE OF LIGHT.—It is noteworthy that periodic stars generally, if not universally, take less time in the increase than in the diminution of their light, and in some instances they vary these times of increase and decrease. Delta Cephei takes one day and fifteen hours to pass from its minimum to its maximum, and three days and eighteen hours to pass back again to its minimum. Zeta Geminorum takes four days and twenty-one hours for its increase, and five days and six hours for its decrease. Betelguese take ninety-one and one-third days for its increase, and one hundred and four and one-half for its decrease. Mira Ceti on a mean takes fifty days to increase from the sixth magnitude to its maximum, and sixty-nine days to decrease again to the same magnitude; so that this star is visible to the naked eye for about four months on an average; occasionally for five months, and at times for only three months. Of course the periods of its increase and decrease vary accordingly, and at times the former is longer than the latter. The shortest observed period of increase was eighty days; this was in 1679: the longest was sixty-seven days; this was in 1707. The longest period of decrease was ninety-one days; this occurred in 1839: the shortest was fifty-two days; this took place in 1660. The constancy and the inconstancy of the periods of increase or decrease may be accounted for in the same way as the constancy and inconstancy of the routines of the changes in the irregularities already described. In fact, the changes in the routines of irregularity, and the changes in the length of time for the increase or decrease of light, are only different views or names of the same thing.

IRREGULARITIES IN THE PERIODS.—We are now prepared to understand the variations in the periods of the periodic stars. It has been made evident that variable stars have a regular and an irregular element. The regular element is their invariable rotation on their axes, the irregular element lies in the changes of the sizes and positions of their spots. The punctual periodicity of some of these stars may be readily accounted for by their turning round always with the same rate of motion, and by the invariableness in the sizes and positions of the spots; and all the various irregularities we have just described, and still others to be given, may be accounted for by irregularities of the spots similar to those with which we are so familiar in the sun. The period of the sun's rotation on its axis is inferred from our observations of its spots, and sometimes that period seems much longer than at others. This occurs because the spots are rapidly moving about, and hence there is an uncertainty of forty-eight hours or more about the real period of the sun's rotation. Judging from our observations on one spot alone, we would infer that the sun rotated once on its axis in less time by two days, than from our observations on some other spot. The sizes of the sun-spots vary as wonderfully as their positions: some have been known to contract their diameters at the rate of more than a thousand miles a day. In this way we see the identity between the irregularity of the sun and the irregularity of the fixed stars.

We know with absolute certainty what causes the irregularity of the one, and it is plain, therefore, what causes the irregularity of the other. The sun and the fixed stars are bodies of the same nature. From their masses, as seen in the binary systems, and from the amounts of light coupled with their distances, we know that they are similar in their sizes, though having diversities among them as among the planets of the solar system. These irregularities in the periods of the fixed stars are, in truth, new bonds of connection between distant parts of the universe: they show in a new aspect the intimate family likeness between the sun and the stars.

There is another way of connecting the irregular periods of the periodic stars with the changes in the sizes and positions of the spots. These changes in the spots, as we have seen, produce changes in the routines of irregularity in the light of the stars. Spots when large and irregularly disposed, cause irregularities in a star's light. When the sizes or positions of the spots are changed, then the irregularities in a star's light must change correspondingly; and with this change of routine the change of the star's period is connected by the following law:

THE GREATER THE IRREGULARITIES IN THE ROUTINES OF LIGHT FROM ANY PERIODIC STAR, THE GREATER ARE THE IRREGULARITIES OF ITS PERIOD: WHEN THE ROUTINE OF IRREGULARITIES REMAINS THE SAME, THEN THE PERIOD REMAINS THE SAME. This law may be illustrated by many facts, and it shows that *the changes in the periods of the periodic stars, are due to the same cause as the changes in their routines of light.* If the changes of the spots produce the one, they must produce the other. First, we will review the instances where the routines and the periods are constant, and then the instances where both are alike inconstant.

Algol in 1784 had a period of two days, twenty-one

hours, forty-eight minutes, fifty-nine and a half seconds: in the year 1842, that is, fifty-eight years afterwards, the period was only four and one-fifth seconds shorter. Careful observations now going on will probably show that this small difference is caused by changes in the distance between that star and the sun. As four and one-fifth seconds represent a distance travelled by light less than a million of miles, it will at once be seen that so short a difference in the period may have been made by a change of distance. The period of this star, therefore, may be regarded as regular; its routine of changes also remains the same.

Delta Cephei. "Of all known variable stars, this exhibits the greatest regularity *in every respect*," that is, both in the changes of its light and the duration of its period.

Eta Aquilæ. "At long intervals of time trifling fluctuations occur in the period of this star, not amounting to more than twenty seconds," and these may yet be found due to changes of distance. Its changes of light are as constant as its period is regular.

Zeta Geminorum. "This star has hitherto exhibited a perfectly regular course in the variations of its light," as well as in its period.

Beta Lyræ. This star proceeds with great constancy through its routine of two maxima and two minima, in a very regular period. In 1784, its period was shorter by two and a half hours, and in 1818 by one hour, than in 1844, when it was twelve days, twenty-one hours, forty-eight minutes, forty seconds. In 1850, the shortening was again clearly perceptible. This regularity in increasing and decreasing its period will probably be found due to regular changes of distance.

In these five stars the constancy of the periods is truly wonderful, and the routines in their light are just as constant; both facts, doubtless, being alike due to the same cause—the steadiness of the spots. We may now briefly review other stars whose periods are irregular, and whose routines in their changes of light are inconstant, both alike to be accounted for by the changes of the spots:

Scuti R sometimes varies from the fifth to the ninth magnitude; at other times it varies less than a single magnitude. "The duration of its period is also subject to considerable fluctuations."

Alpha Cassiopeiæ. "The difference between its maximum and minimum is small, and is, moreover, as variable as the duration of its period."

Virginis R "maintains its period and its maximum brightness with tolerable regularity; some deviations, however, do occur."

Leonis R. "The period is somewhat irregular. The brightness of the maximum seems also to fluctuate."

Alpha Herculis. "Frequently its light scarcely changes for months together; at other times in the maximum it is half a magnitude brighter than in the minimum;" consequently the period also is very uncertain.

Coronæ R "is variable only at times," and its period is difficult to determine.

Mira Ceti. I have already given a statement of the great inconstancy of this star in the routines of its increase and decrease of light. Sometimes it is visible five months and sometimes only three. In one rotation it requires only thirty days, and in another nearly seventy days, to increase its light. In some rotations it takes fifty days, and in others ninety days to decrease its light. These numbers refer to the times when it is visible to the naked eye. Its entire period is equally irregular: it has been as short as three hundred and six days, and as long as three hundred and sixty-seven days. Between these two points it varies by degrees, but not regularly, differing occasionally as much as twenty-five days from any law of change that can be made to apply." * No changes of distance can explain the large and irregular variations of its period. These must have the same cause as the variations of its routines of light, namely the changes of its spots.

Chi Cygni. This star, in the irregularity both of its routines and period, is much like Mira Ceti.

30. Hydra Hevelii. "Both its period and its maximum brightness are subject to very great irregularities."

Epsilon Aurigæ. The fluctuations of its light are so irregular that its period has not been determined.

Instances enough have now been given. Among all the other periodic stars within my knowledge not here mentioned, there are none which differ in character from these, or point to any other cause of irregularity. The law above announced seems fully established, and with it some important consequences.

We may assume that a spot on the sun 45,000 miles in diameter intercepts much light. Occasionally there are close groups of spots 120,000 miles in diameter; and more than one hundred spots, great and small, are visible at a time, the smallest of which are 1,000 miles wide. Around the spots, however, there is generally a superior brightness which acts partly as a compensation. Had we a photometer as delicate in measuring light as are our thermometers in measuring heat, doubtless we would see that these great spots and groups of spots, lessen the light of the sun. Assuming that they do, we have this statement of fact and inference. The sun by his spots is made a periodic star. By the changes of the positions of the spots, his periods are made irregular, differing from one another as much as two days. By the changes in the sizes of the spots there is an

* "Cosmos," vol. iii., p. 229. Note.

irregularity in his maxima and minima, and in the routine of his changes. In the fixed stars we see *the same connection between the irregularities of the periods and the irregularities of maxima and minima and of their routines.* These two connected irregularities in so many fixed stars must be due to the same cause as the like connected irregularities in the sun, namely, to the changes both in the sizes and in the positions of the spots. The connection of two such distinct irregularities in the stars, coupled with a like connection of distinct irregularities in the sun, is a weighty consideration. It lends additional and independent evidence for the existence and influence of spots in the stars.

But what shall we say of those periodic stars whose periods and routines of light do not change? We have seen that the constitution of the sun is liquid, and that his spots are cooled and hardened portions floating in the rapid currents—they are floating islands, in fact. But in the case of periodic stars whose periods are invariable, the spots do not move. They must then be continental in their characters; they reach more or less around the star, and form a firm, immovable framework, allowing still vast floods of light to flow forth from those parts which are not yet covered by the spots. The crust of our earth is believed to rest on a liquid ocean below. Once the solid parts were only floating islands; then these as they grew, coalesced forming one body which moved about no more; but this body was not yet large enough to cover all the liquid beneath, or to obstruct all the light. Our earth was then a periodic star, with a regular period and a constant routine of light changes.

On well-known physical principles, and in accordance with familiar facts, we have now seen an easy explanation for all the peculiar phenomena of periodic stars, save one, and that is the shorter time required for the increase than for the decrease of their light. If the interception of their light be made by spots in the form of floating islands, then the latter end of an island, as the star moves round, must be broader than the preceding end. If the figure A B represent a part of the surface of a star, and C D 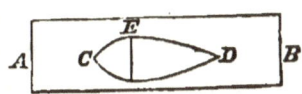 a spot, then as the rotation is from A toward B, the end D of the spot would first appear, and the light would diminish until it had passed as far as E, the widest part representing the minimum of light. Then the augmentation of light would begin, and plainly it would require a shorter time for the increase of the light, because that end of the spot is shorter. From recent very imperfect researches it appears to me that the spots and the groups of spots in the sun move with their longer and narrower ends foremost. This, however, needs confirmation. If it prove true, then we have the very explanation demanded. But why in a group of spots should there be this tendency in the broader ones to lag behind? One reason seems to be this: The largest spots would be the thickest, and if so would float highest above the general liquid surface. Solid lava floats buoyantly on liquid lava, as seen in volcanic streams and pools: solid slag floats lightly on melted slag at an iron furnace: solid cast iron swims easily on cast iron melted. The higher the island the slower it would move, and acquire more and more a tendency to lag behind. This is one of the laws of rotary motion: if our earth were to expand, it would rotate more slowly; if it should contract, its rotation would be hastened. Moreover, sound reasons have been given for currents in the atmosphere of the sun in the equatorial regions, moving in a direction contrary to the sun's rotation, similar to the trade winds on our earth. For the same reasons currents of the same kind must move in the atmospheres of the stars, and hence the largest floating island of a group, or the largest end of a spot, from its height and breadth, would be delayed by the atmospheric current more than the smaller ones, and would thus be placed in the rear. Thus the fact that a shorter time is required for the increase than for the decrease of a star's light, confirms the idea of floating islands.

We may derive an additional confirmation from viewing a terrestrial globe, a picture of our planet. There are weighty geological reasons for believing that the continents and large islands represent those portions of the earth's crust which first solidified, and the oceans those portions which remained liquid the longest; and that when general contraction and wrinkling took place, the softer, thinner, and more recently hardened portions sank down, forming the ocean floors. The continents and islands, therefore, would represent in a general way the spots. Then with such spots on our burning and shining globe, the maximum of light would be over the Pacific Ocean; slowly there would be an *irregular* decrease of light, and the minimum would be a little east of the centre of Africa. Then there would be a much shorter time for the increase of light and a second maximum over the Atlantic Ocean. The second minimum over the American continent would have nearly equal times for the decrease and the increase of light. All the changes would be irregular, as in the periodic stars; there would be no steady rate of augmentation and diminution of light. And there would be two maxima and two minima, as in the case of Beta Lyræ. If the ocean were lifted up as it was formerly in a state of vapor by heat, we would see the continents united at Behring's Strait, where the water is now quite shallow. They thus formed a firm framework, preventing irregular movements. To the eastward of the largest continent there lie scattered for a long distance very many islands. The atols prove that much the largest amount of them have sunk slowly down beneath the present ocean level, as was shown by Darwin. How remarkably this coincides with the idea already mentioned, that the smallest spots of a group generally float before! In front of the American continent the same fact appears again.

It was recently proved that the irregularities in the periods of the periodic stars, formed a bond of connection between them and the sun. Now it appears that the perfect regularity in the periods of other periodic stars, connects them with our earth. It may be said that all this is pushing a theory too far. On the contrary, the true and ultimate test of any theory is to apply it as closely as possible in all the details of its consequences. This strengthens the right theory, and invariably breaks down a wrong one.

Another recommendation of this theory is, that it ignores all essential distinction in their natures between the fixed stars, the sun, the earth, the moon, and the other planets and satellites. It regards them as bodies all fundamentally the same, but in different stages of the same progressive change. This same truth is strongly made evident by other independent astronomical evidence. The classification into sun, planets, and satellites, is useful merely to show the relations between the members of our own, or of other similar systems; but I have already proved that this distinction fades away when we approach other systems of double, triple, and multiple stars. Often it is impossible to say which is the sun or the planet, or to decide which is the planet and which the satellite.

It has been supposed that periodic stars are caused by dark planetary bodies revolving around them, and thus periodically obstructing their light. This idea would apply best to Algol, because the interruption of its rays is for a shorter time than in any other star. Still this interruption of more or less of its light, amounts to about one-eighth of its entire period, and hence the planetary body would have to be so large as to occupy about one-eighth of its own orbit; this is quite an impossible condition, for the two globes of such a system would be at once in contact. A planetary globe would obstruct the light of a star in waning and waxing with mathematical regularity. But I have tried to show prominently that the interference of their light goes on with all imaginable routines of irregularity, and that often in the same star these routines alter.

IRREGULAR STARS.

Cases occur when it is doubtful whether a star should be classed among the periodic or the irregular stars, as for instance Hydra Hevelii and Epsilon Aurigæ. We can easily understand how this may be. If the period of the rotation of the star be long, from three hundred to five hundred days, and if the changes in the sizes and positions of the spots be extensive and rapid, then the variations of the light arising from the variations of the spots may entirely obscure and overcome any signs of the star's rotation. The irregular element may be much more conspicuous than the regular. The rotation may go on, and be punctual to a second, but the spots may not be permanent enough to indicate any rotation at all. There may be still another cause for irregular stars. When treating of the sun and of the earth, we saw many proofs that the light and heat of the sun and stars are caused by chemical action. This action is generally steady, not varying perceptibly in amount; but because it is generally steady, that is no reason why it should be always so. We know of nothing to hinder occasional unsteadiness, or even extraordinary fluctuations in its energy. Therefore, if we are convinced from sufficient reasons that chemical action causes the light and heat of the stars, and if we see some stars very variable in the amount of their light, then we are at liberty, in the absence of all other causes, to conclude that chemical action in the stars, as well as here on the earth, is subject to obstructions and also to extraordinary activity.

Coronæ R has been called a periodic star with a period of three hundred and twenty-three days, and, as already stated, it is variable only at times. But its irregularities are most conspicuous. In the winter of 1755–'56 it became totally invisible; subsequently it again appeared, and the variations of its light were observed as given in the table. In 1817 its brightness was nearly constant, and in 1824 it again became variable. Its constancy again returned from 1843 to September, 1845. Toward the end of that month a fresh diminution of its light commenced. By October it was no longer visible to the comet-seeker. but it reappeared in February, 1846, and by the beginning of June had reached its usual magnitude, the sixth. From that time until 1850 it was invariable, or nearly so.

Eta Argus in 1677 was of the fourth magnitude, and by 1751 it was of the second. In 1815 it was of the fourth. In 1826 it was of the second. In 1827 it was of the first, and declined the same year to the second. In 1834 it was between the first and second. In 1837 it exceeded all the stars of the first magnitude, except Canopus and Sirius. In January, 1838, it again became slightly fainter, until March, 1843, still remaining of the first magnitude. In April, 1843, it again began to increase, and was superior to Canopus and almost equal to Sirius. Since then it has lost much of its brightness, and gone through various changes. Such irregularities can be accounted for only by the supposition of irregular action in that process which produces the light of the stars. If that process be chemical action, there is nothing strange or incredible in its varying activity.

LOST STARS.

Many stars are marked on the maps which no longer shine in the heavens. Some of these may have been planets which have moved out of their recorded places, and others may have been errors of position on the maps. But after making all due allowances, it remains certain that some stars have entirely disappeared.* This is very natural, and just what we should have expected from the fact that it is the nature of chemical action in any body to come to an end. If the cause of the light and heat of the sun and stars be chemical action, then lost stars, periodic stars, and irregular stars, are quite as much in the regular course of things as the changing and falling leaves of autumn; they are, in fact, an absolute necessity. In the composition of our globe we see all the elements chemically united, the same as they must necessarily be combined by a great burning. The slight exceptions of the isolation of some elements may be accounted for. Hence the former fires of our planet are extinct; their heat alone remains, and our globe may be regarded as a lost star. It shines no more by its own independent light, but, like Venus and Jupiter, it glows mildly with the borrowed light of the sun. Laplace, speaking of temporary stars, said: "Those stars that have become invisible after having surpassed the brilliancy of Jupiter, have not changed their places. There exists, therefore, in celestial spaces dark bodies of equal magnitudes, and probably in as great numbers as the stars." Bessell said: "No reason exists for considering luminosity an essential property of these bodies. The fact that numberless stars are visible is evidently no proof against the existence of an equally incalculable number of invisible ones." As lost stars are

* Herschel's "Outlines."

a necessity for the theory of chemical action, they afford strong support to that theory.

TEMPORARY STARS.

In order to show the cause of temporary stars, I will briefly describe them, giving only such items respecting them as bear on our purpose, leaving out all historical sketches relating to their discovery and contemporary events. The Chinese records contain accounts of several temporary stars, but they are too deficient and unreliable in particulars to be of service here. Each of the following dates, at the beginning of the paragraphs, represents a new temporary star:

134 B. C. In the constellation Scorpio, and seen by the Greek astronomer Hipparchus: it was a large, conspicuous star.

386 A. D. In Sagittarius: it equalled Venus in brilliancy, and was seen only about three weeks.

827. In Scorpio: recorded by the Arabian astronomers as "equal in light to the moon in her quarters."

945. Between Cepheus and Cassiopeia: it must have been large to have attracted attention in these dark ages.

1012. In Aries, and of extraordinary magnitude, dazzling the eyes, increasing and decreasing in size, and becoming sometimes even extinct. It was seen three months.

1264. Between Cepheus and Cassiopeia: it must have been large to have drawn attention in such an age.

1572. On the 6th of August, between Cepheus and Cassiopeia: it is suspected to have been the same as the last, and also as the star of 945 A. D. Its brightness at first was greater than that of Sirius or Jupiter, comparable only to Venus, and discernible by keen eyes in a clear air in the daytime even at noon; at night also through the clouds, when not very dense, and when all the other stars were hidden. It scintillated more strongly than stars of the first magnitude. In December it began to diminish, and it decreased gradually until the end of nineteen months after its first appearance, when it vanished. Its changes of color will be given among the colored stars.

1600. In Cygnus, of the third magnitude: after remaining several years it began to decrease in brilliancy, especially after the year 1619, and it vanished in 1621. It was seen again in 1655, of the third magnitude, and again it disappeared. In November, 1665, it appeared again, extremely small, then larger, but not reaching the third magnitude. Between 1677 and 1682 it decreased to the sixth magnitude, and so it still remains. Sir John Herschel classes it among the variable stars, in which he differs from Argelander.

1604, October 10th. In Ophiuchus: it was of the first magnitude, greater than Jupiter, but less than Venus; it gradually decreased, and disappeared after about seventeen months.

1670, June 20th. In Vulpes: it was of the first magnitude, but by the 10th of August it had diminished to the fifth; it disappeared after three months, but reappeared again March 17, 1671, when it was of the fourth. In April its brightness was very variable. It recovered its original splendor in ten months, but in February, 1672, it disappeared. It reappeared again, March 29th, as a star of the sixth magnitude, and since then it has not been seen.

1848, April 28th. In Ophiuchus of the fifth magnitude; and in 1850 it decreased to the tenth.

By this brief statement of facts it appears that temporary stars have been eminently variable in their amounts of light, and therefore they have shown their intimate relationship to the class of irregular stars, especially to those which have been most irregular, as Eta Argus and Coronæ R.

Three of them disappeared and appeared again, two of them repeating this operation twice. At their first appearance they all, except three, burst forth with the effulgency of first-magnitude stars, some of them more brightly than any other star. This large blaze of light, their disappearance so soon, their extreme variability during their short existence, even to disappearing and appearing again, show that they are an exaggeration of the character of Eta Argus, which once rivalled the brilliancy of Sirius. Their permanent disappearance connects them with lost stars. Their changes of color connect them with colored stars, especially to such as change their colors. Therefore, as they are associated indissolubly with irregular stars, with lost stars, and with colored stars, they do not stand alone as distinct phenomena to be accounted for only by an independent cause; but the same cause which produces any one of these classes, must produce them all.

The theory of chemical action is equally adequate for the explanation of all these classes. This action may go on so feebly, that from such vast distances we may be unable to see its light; or it may proceed with such unwonted energy, that even Sirius may be outdone. We do not know what limits to set to its widest extremes of variation; but we do know that it may vary, and often does vary greatly. The causes or conditions of its variation are numerous, and each individual case has its peculiarities; and knowing so little about the stars, we cannot enter into the particulars. Pseudo-volcanoes on our earth may be an illustration of the general principle. The original carbon of our planet entered into combination with oxygen "in the beginning," and formed carbonic acid, which made a part of the atmosphere and floated around the globe. By a very long process, seemingly incredible at first, this same carbon was separated from the oxygen, and deposited in the form of coal and petroleum in the ground. There was also deposited, in connection with this, a combination of sulphur and iron. By contact between this sulphuret of iron and the atmosphere, a new chemical process takes place which results in a pseudo-volcano; all the carbon in the vicinity of the sulphuret of iron begins once more to combine with the oxygen of the air, and heat and light on a grand scale are produced. Such events occur in the western part of the United States, beyond the Mississippi. Sulphur may be further described as another example. In the early history of the formation of our globe all the sulphur combined with other elements. But recently in volcanic operations these sulphurous combinations are slowly decomposed, and the sulphur is set free again to combine with the oxygen of the air and with other elements. Such chemical events as these just mentioned in the history of sulphur and carbon, could not have been anticipated by any reasoning *à priori*. Neither can we anticipate how, in the stars, long and intricate processes shall take place, decomposing old combinations, and forming new ones on a grand scale, that shall produce heat and light visible even here. Spontaneous combustions on our earth are not uncommon. Buildings on the land and ships on the ocean are often wrapped in flames by these mysterious agencies. What occurs on a small scale may occur on a large one. All are familiar with the irregular flickering of a dying lamp. After the extinguishment of a fire in a burning building, the flames not unfrequently burst out again and spread a new alarm.

None of the chemical elements are more combustible in ordinary conditions than phosphorus. Yet even phosphorus, after burning a while, goes out, and then it may be rekindled to burn vividly again. The experiment is beautifully performed by taking a piece about a quarter or half an inch in diameter, placing it on a slate or smooth stone, and touching it with a match. While burning it will melt, spread over the slate, and after a while expire. Scrape together the residue with a knife-blade, and in so doing it will rekindle. This illustrates a principle of grand importance in our present inquiry. It is this: all burning bodies, small and great, must soon be extinguished by their own incombustible products, unless these products be removed by some external agency. A lamp would expire in a short time if the surrounding air had not been so contrived as to originate currents to carry off the products of combustion. This principle is applicable to chemical action everywhere, on the earth, in the sun, and in the stars. Hence we should expect that if chemical action, in the form of combustion, be the cause of light and heat in the stars, then, among hundreds of thousands of cases, some few must occur where that combustion is liable to be extinguished by its own incombustible products, or so nearly extinguished that in our distant planet we may no longer see its light. But such incombustible obstructions may, after a longer or shorter time be removed, though it would be rash for us to say exactly how. They may ultimately be removed, as here, by currents in the atmosphere, or by currents in the liquid body of the star; or they may be made to sink down out of the way; or they may be brought by either one or both these kinds of currents into contact with other substances, which may cause them to burn again. Our great ocean, for instance, is the incombustible product of a great conflagration, but it may all be burned over again by simple contact with a sufficient amount of potassium; and in this case, very much to our purpose when we come to consider changes of color among stars, the second burning would be with a flame very differently colored. The removal or the renewal of these incombustibles may require a short time, or millions of years, as the renewal of our carbon in the forms of coal and wood from the carbonic acid in the air.

Such facts and chemical laws throw light on chemical action among the stars. It may go on regularly and steadily, or irregularly and unexpectedly strange, even to an appearance of extinction, for incalculable periods. Carbon may burn and pass into an incombustible compound, and after millions of years be separated and burn again. The same is true of sulphur, of oxygen, and of hydrogen. The highly inflammable phosphorus may go out, and again revive with its usual splendor. The flickering of a lamp, the many accidental forms of spontaneous combustion, and the pseudo-volcanoes in the crust of the earth, show the influence of changing conditions on chemical action. Among the incalculable number of stars, and therefore infinite varieties of conditions, it is more natural for us to look for a few cases of irregularity in their chemical actions than for a uniform sameness. We must necessarily, therefore, expect irregular stars and temporary stars. Their deep wonder and their mystery vanish before the theory of chemical action, and thus they afford a firm support for that theory.

COLORED STARS.

The stars present two great facts in regard to color. The first is the differences between their colors, which extend to the widest degree; the second is their changes of these colors. The theory of chemical action, in its natural course and without any violence, explains both these phenomena. Our artificial lights and fires are generally of the same color, because they are mainly the products of a combination between the same substances, namely, oxygen on the one side, and carbon and hydrogen on the other. This is the case with gas, tallow, wax, oil, spermaceti, wood, turf, and coal of all kinds. But when we experiment on all the inorganic elements and their compounds in a state of combustion, they exhibit the widest possible range of colors, and various tints and shades of the same color. Sulphur, zinc, potassium, and sodium thrown on the surface of water, brass wire that has been dipped in acids, and all the beautiful displays of pyrotechny, are familiar examples. The pleasing lilac-purple flame of cyanogen is always remembered. As already observed, the sun consists of many thousands of simple chemical elements, and the stars evidently have similar constitutions. On the theory that these elements are in a process of burning, they should display, like pyrotechny, all imaginable beautiful colors. And this is what they do. But when only a faint beam of any colored light strikes the eye, it gives no sensation of color different from white. The large stars, the seventeen of the first magnitude, for example, are all, when carefully examined, of various colors. They are collected in a tabular form a few pages forward, for the purpose of giving a connected view not only of their present, but also of their past colors, as far as yet known.

When speaking of them generally as red, blue, and the like, the precise tints of these colors are omitted; but the shades of red in Aldebaran, Betelguese, and Antares, differ sensibly.

Miss Maria Mitchell has given a careful statement of the different shades in the colors of thirty-six double stars in the number for July, 1863, of the "American Journal of Science and Art." She selected them, not for their colors, but for the measurement of their distances apart, and their angles of position. Thirty of the number are particularly described as variously colored; five were observed in weather unfavorable for observations on color, and of one nothing is said. This is an important announcement: of thirty-six stars selected at random, at least thirty are colored, and probably more. Differing as it does from our ordinary ideas of the colors of stars, still I presume from theoretical considerations it exhibits a fine sample of what we shall have of all the larger stars, when this department of astronomy shall be well cultivated. The extreme smaller orders will be mostly represented as white, because their light is too feeble to give the impression of color.

Struve, in his great catalogue of 2,787 double stars, published in 1837, speaks particularly of the colors of six hundred of the largest. The majority of these six hundred were colored other than white in either one or both the companions. Of those which he records as white, the larger number probably will hereafter be found to have delicate shades of various hues when carefully examined with colors as the prime object in view. Their distances and angles of position were his chief object, and nice varieties of tints were not in that day much regarded.

Thus from these three statements, one concerning the stars of the first magnitude, one respecting thirty-six double stars, and a third regarding six hundred double stars, we have the evidence that when the stars are large enough to give the impression of color either by the naked eye or by the telescope, they have all that prevalent variety of hues which the burning of mineral elements displays. This is another powerful confirmation of the chemical theory of the light of the stars. We must not look for the strong and vivid effulgence of pyrotechny, because the amount of starlight is so small; as we find, for instance, when we attempt to read or to do aught else by starlight.

In such bodies as the sun and the fixed stars, it is not to be supposed that all the elements can occupy the surface and give out light at the same time. Only a few probably can do this at once, and the united colors of these elements give the color for the time being to the star. Nor is it in the nature of things that the same elements should always be the light-giving ones at the surface. The very process of chemical combination forbids this. When any one element is fully combined, then its light-giving power ceases, until it shall be decomposed and released again, to run a new round of combination. In the mean time another element, or set of elements, enjoys the predominance of combustion at the surface, and, of course, of giving out light. The light of this last set may be of a different color from the former, and hence the star would change its color. In strict accordance with this very natural view, we find that the stars do actually change their colors.

When from theoretical considerations I believed that the stars change their colors, I began to search for the evidences of such actual changes. But not a word is said about them in any work on astronomy in the English language. Arago in his French work "Astronomie Populaire," gives a few sentences, and these relate chiefly to discrepancies in color between the observations of the elder Herschel and the elder Struve on the double stars. I therefore began a course of observations of my own on the larger stars, all those of the first and a few of the second magnitude. These were selected because their colors can be unmistakably seen by the naked eye on clear nights, at any time, even when walking along the streets. By selecting them we can enjoy the advantage of asking any friends casually present their opinions about the color of a star, when we wish our own opinions confirmed, or, if need be, rectified. I have received valuable assistance in this way. At the same time I endeavored to find in books, journals, or periodicals, any scattered casual notices about the colors of stars. These researches led me to make five short communications to the

Academy of Natural Sciences in this city; extracts from which, properly abbreviated and altered, I will here present:

Feb. 10, 1863.—During the past year I have made the colored, the variable, the periodic, the lost, and the temporary stars a special study. Catalogues have been made of some of these classes of stars, but no catalogue has yet been made of stars which have changed their colors. Indeed, Humboldt, in writing about the red color ascribed to Sirius by the ancient Greeks, says, "Sirius, therefore, affords the only example of an historically proved change of color, for it has at present (1850?) a perfectly white light." And yet, in apparent contradiction to this, he, in other pages of the third volume of his 'Cosmos,' mentions other fixed stars whose colors in modern times have been known to change.

This change of color is one of the greatest physical events. Think of our own intensely-lighted sun, 2,770,000 miles in circumference, as being deeply red, then turning to be perfectly white, then changing to purple, and then again to green! What mighty causes must be in operation to produce such grand results! This should be made a distinct section of astronomical study, and allied to that of the other classes of stars just mentioned; and I therefore offer the following as an enumeration of stars whose colors have changed.

CATALOGUE.

1. The temporary star of.1572, which suddenly broke forth with such splendor that it could be seen with the naked eye by day at noon, and by night through the clouds when all other stars were hidden, shone for the first two months with a white light, then it changed to yellow, then it assumed the deep red of Mars, then it appeared with the

lighter red of Betelguese, then it took on a shade of red like that of Aldebaran, and at last when quite small it shone with a dull whiteness like that of Saturn. Such changes in color, along with changes in amount of light, seem natural in a process of combustion among many mineral elements, especially in a waning state, when one element after another was expiring.

2. Eta Argus.—Humboldt, in the third volume of his Cosmos, says that this star " is undergoing changes in color, as well as in intensity of light. In the year 1843, Mr. Mackay noticed at Calcutta that this star was similar in color to Arcturus, and was therefore reddish yellow; but in letters from Santiago de Chili, in February, 1850, Lieutenant Gilliss speaks of it as being of a darker color than Mars."

3. Beta Ursæ Minoris has been described by the careful German observer Heis, in these words: " I have had frequent opportunities of convincing myself that the color of this star is not always equally red; at times it is more or less yellow, at others most decidedly red."

4. Alpha Crucis.—Humboldt says: " My old friend, Captain Berard, who is an admirable observer, wrote from Madagascar, in 1847, that he had for some years seen this star growing red."

5. Capella.—Ptolemy associated in color the red star Mars with Capella. In 1850 Humboldt wrote that El Fergani, in the middle of the tenth century, on the Euphrates, had " described as red stars Aldebaran, and, singularly enough, Capella, which is now yellow, with scarce a tinge of red." " Singularly enough," comes out naturally from that veteran philosopher, at a time when changes in the colors of stars were regarded as incredible. " Riccioli, in the 'Almagestum Novum,' edition 1650, also reckons Capella, together with Antares, Aldebaran, and Arcturus, among red stars."— " Cosmos," vol. iii. In September, 1859, the Rev. J. B.

Kearney, in a letter to Sir J. Herschel, printed in the 20th volume of the "Monthly Notices of the Royal Astronomical Society," says, "By the way, the color of Capella seems much less blue than it used to be." To myself, at present, February, 1863, its color appears to be a delicate pale blue. Capella, therefore, is recorded to have had four colors—red, yellow, blue, and a shade "much less blue."

6. Sirius.—In the times of the old Greeks and Romans, Sirius was always spoken of in strong terms as being red; Seneca says, "redder than Mars." In the middle ages the Arabian astronomers did not name Sirius among the red stars, neither did the earlier astronomers of the west of Europe. Therefore it seems probable that its color changed from red to white, between the times of the Roman and those of the Arabian observers. Humboldt, in 1850, says, "it is perfectly white." In 1860, in this city, Dr. Wilcocks and a lady friend of his who was qualified to distinguish nice shades of color, pronounced the light of Sirius to be purple or violet. In October, 1862, it appeared to myself and some friends as the most deeply colored star in the sky, but as more green than blue. Ever since then it has been decidedly green, though during the autumn of 1863 and the beginning of the year 1864 the green was paler. The colors of Sirius, therefore, have been red, "perfectly white," violet, and green.

Catalogue continued, March, 1863.—7. Procyon.—In 1850 Humboldt classed Procyon among the yellow stars. Donati, in 1862, quoting Schmidt without date, gives this star as being white. Feb. 17, 1863, I announced to this Academy that Procyon is very decidedly blue; and in this all to whom I have referred the color, agree without hesitation.

8. Rigel.—This star is classed among the white stars by Donati, in a Memoir dated August, 1860, and published in the Annals of the Museum at Florence in 1862. It is

now decidedly blue. During the past two months it has been observed by myself and some friends to be one of the most deeply colored of all the stars now visible in this latitude.

9. Alpha Lyræ, or Vega.—Donati, in the Memoir just named, classes Vega among the white stars. Humboldt, in 1850—"Cosmos," vol. iii., p. 183—says, "the light of Alpha Lyræ is bluish." To myself it now appears pale blue, very much like Capella.

10. Castor.—Donati, in 1860, classed Castor among the yellow stars. Humboldt, in 1850, says, " Castor is a greenish star."—" Cosmos," vol. iii., p. 177. It appears to me greenish now—March, 1863.

The two companions of this double star have been further described as follows: yellow and yellowish by Sestini; greenish yellow and green by Dembowski; bright and pale white by Webb; yellow and warm yellow by Miss Maria Mitchell.

There is a close cluster of more than a hundred stars, known as Kappa Crucis, in the southern hemisphere, and when seen through a telescope, the very varied colors of its individual stars give it, according to Sir John Herschel, " the effect of a superb piece of fancy jewelry." During his residence at the Cape of Good Hope, he made a drawing or map of this group, and stated carefully the colors of eight of its most conspicuously colored stars. Just a quarter of a century later, F. Abbot, Esq., in a communication to the Royal Astronomical Society in England, dated, Private Observatory, Hobart Town, May, 1862, describes how this piece of jewelry has changed. Six of these eight stars have now different colors. The changes, according to him, are as follows:

11. Gamma: changed from greenish white to bluish purple. There is an error in Mr. Abbot's communication,

as printed in the "Monthly Notices of the Royal Astronomical Society." The name of this star is there printed Nu, instead of Gamma. Sir John's list has no Nu.

12. Delta: has changed from green to pale cobalt.
13. Epsilon: has changed from red to Indian red.
14. Zeta: has changed from green to ultra-marine.
15. Phi: has changed from blue green to emerald green.
16. Alpha[2]: has changed from ruddy to the similar color of all the small stars of that magnitude. "The smaller stars, from the tenth to the fourteenth magnitude, are generalized, and all partake of nearly the same color—Prussian blue—some with a little more or less tint of red or green mixed with the blue."

It is to me an impressive fact that so many conspicuous stars now nightly appear blue or green; especially as the first blue stars were mentioned by Mariotte, so lately as in 1686;—before him, no departure from white was named but red, with the exception of the yellow in the temporary star of 1572. Capella, Bellatrix, Rigel, Procyon, Vega, Spica, are blue; all deeply so except Capella and Vega. Sirius is conspicuously green, and Castor slightly green, though I sometimes doubt the greenness of the latter. When a star is not of the first magnitude, or when its departure from white is not very decided, a long fixedness of the eye upon it is necessary, and a careful exclusion of artificial lights. At least, this is my experience. It has occurred to myself, and been suggested by others, that perhaps this blueness of so many large stars now in view, and greenness, which is a modification of blue, may be owing to some special atmospheric cause. For many months the air has been unusually damp. But this cannot be the cause of these colors, for Aldebaran, Betelguese, and the planet Mars are in the midst or near vicinity of some of these blue and green stars, and preserve their ordinary redness. Pollux, however, in the same general neighborhood, seems hardly entitled to be called a red star. From the fact that it was classed among the six decidedly red stars by the ancients, I regard it as changed, but desire further time before announcing the change in the catalogue. Humboldt calls it merely "reddish."

I would not be premature in speculating on the laws or causes of these changes, but must remark that the change of Sirius, from its ancient red to white, and now to green, is all in the same direction, namely, a *relative* diminution of the red. The three primary colors, red, yellow, and blue, with an excess of red, will give a red star; take away the excess of red, and the star will be white; take away still more of the red, and it will be green—that is, a combination of the remaining yellow and blue. The three colors of Sirius, therefore, the ancient red, the modern white, and the recent green, may possibly be due to the operation of a single cause. With the ordinary proportions of red and blue in a white star, a slight diminution of the yellow would make the star purple or violet. Thus the "perfectly white" of Sirius might be followed by the "violet," as next afterwards it was observed.

Catalogue continued, June, 1863.—17. Spica.—This star was described by Humboldt, in 1850, as being "decidedly white."—" Cosmos," vol. iii., p. 181. Donati also arranges Spica among the white stars. It is now conspicuously blue, and has been observed of this color by myself and several friends since early in March.

18. Altair.—Humboldt, in 1850, classes this star along with three others which he says " have a more or less decidedly yellow light." Donati also arranges it among the yellow stars. Altair is now (June, 1863) plainly blue.

19. Deneb, or Alpha Cygni. "Decidedly white," are

the words employed by Humboldt to register this star in 1850. At present it is decidedly blue, and it has been of this color since early in May.

Vega.—This star is placed in No. 9 of this Catalogue. Humboldt, in 1850, described it as bluish; Donati, in 1860, as white, on the authority of Schmidt. During February and March of this year, I often compared Vega and Capella together when they were at the same altitude in the northern sky, and they seemed of the same tint, "a delicate pale blue." An observer in this city, in the middle of May, was very decided in saying to me that Vega was much bluer than Capella. On a reobservation of Vega, and from my recollection of Capella, I assented to this opinion freely. This star, I believed, had deepened its blue. Immediately there recurred to my mind the sentence of Kearney, about Capella, in 1859: "By the way, the color of Capella seems less blue than it used to be." Thus both these stars had changed their blue, though in opposite directions. On the 8th or 9th of June I requested a friend, who is not at all a star-observer, to tell me of what color he regarded that star, pointing to Vega. After looking carefully a sufficient time, he said it was green. I again looked myself, and was surprised to see that it appeared really green. Every night since then I have anxiously watched its appearance, and in very clear nights it seems green, but when the air is vapory or hazy, it seems blue. I have referred it to some half dozen individuals, and they, when the nights have been clear, have also pronounced it of a green color. Last winter, in clear moonlight nights, I often remarked that the green color of Sirius was obscured by the intermingling rays of the moon, reflected from the atmosphere. Now also Vega scarcely appears green by moonlight. There is reason to think that this change in the color of stars from blue to green is not uncommon. Humboldt says, " when

CHANGES OF FIRST MAGNITUDE STARS. 139

forced to compare together the colors of double stars, as reported by several astronomers, it is particularly striking to observe how frequently the companion of a red or orange-colored star is reported by some observers as blue, and by others as green."—" Cosmos," vol. iii., p. 284, note.

STARS OF THE FIRST MAGNITUDE.

Of the seventeen first magnitude stars, the changes of the colors when tabulated, stand as follows: the changes having been in the order they are here placed, the last named being their present colors. The numerals refer to the authorities following:

1. *Visible in this latitude, the 40th degree, whose colors have changed.*

Sirius: red,[1] white,[4,6] violet-blue,[7] green.[7,8]
Capella: red,[1,2,3] yellow,[4,6] deep blue,[5] pale blue.[5,8]
Vega: bluish,[4] white,[6] pale blue,[5] deep blue,[8] green.[8]
Procyon: yellow,[4] white,[6] blue.[8]
Altair: yellow,[4,6] blue.[8]*
Rigel: white,[5] blue.[8]
Spica: white,[4,6] blue.[8]

2. *Visible in this latitude whose colors have not been known to change.*

Aldebaran: red.
Betelguese: red.
Antares: red.
Arcturus: orange yellow.†

All these, along with Sirius and Pollux, were denominated fiery red by the ancients.

3. *Invisible in this latitude whose colors have changed.*

Alpha Crucis: growing red.[9]
Eta Argus; orange yellow,[11] deep red.[10]

* See this Catalogue continued, November 10, 1868.
† See No. 20 of this Catalogue.

4. *Invisible in this latitude.*

Canopus.?
Alpha Eridani.?
Beta Centauri.?
Alpha Centauri. This is a double star, about the colors of the two companions Sir John Herschel says, "Both of a light and ruddy or orange color, though that of the smaller is of a somewhat more sombre and brownish cast."

Authorities. 1. The ancients; Seneca, Ptolemy, etc. 2. El Fergani. 3. Riccioli. 4. Humboldt. 5. Kearney. 6. Donati, quoting Schmidt. 7. Wilcocks. 8. Ennis. 9. Berard. 10. Gilliss. 11. Mackay.

Among the eleven stars of the first magnitude visible in this latitude, seven, according to these evidences, have undergone changes of color, and some of them more changes than one. Among the six stars of the first magnitude in the southern hemisphere, not visible here, two have changed their colors, and of the remainder I can say nothing. And nearly all these changes have been sudden, transpiring in short periods. Moreover, none of the eleven first magnitude stars visible here are white—all are either red, yellow, green, or blue. I look with a great deal of surprise on this tabular statement. Why has it not been made long ago? Probably, in great part, because changes in the colors of stars could not be accounted for by any prevailing scientific theory. It has been rationally assumed that the stars are similar in constitution to the sun, and the sun has been encircled with a theory which affords not the least clue to any changes of color. This theory is most singularly complicated and unfortunate. It surrounds the sun, said to be dark, with an apparatus consisting of five distinct atmospheric envelopes, all regularly arranged one above the other: first, a transparent envelope touching the opaque body of the sun; secondly, an opaque cloudy envelope; thirdly, another transparent envelope; fourthly, another fiery luminous envelope; fifthly, a transparent envelope surrounding all the others. Among such a number of imaginary things, there seems to be no room to imagine how changes of color could occur. Hence the mention of a change of color in a star has been regarded as anomalous, as an inconvenient fact, having no relation to any popular theories, and no appropriate place in the ordinary systems. Hence observations on the colors, and on the changes of colors, have not been stimulated, but rather repressed, by this complex theory of the sun.

Another cause for the delay in this department of astronomy, is the difficulty of deciding on the real colors of the stars. The reason why I did not myself first notice the greenness of Vega, was because I had been accustomed to regard it as blue. I relied unknowingly more on my belief than on my vision. This is the same as when in twilight, or less often in broad day, we think we see an object very distinctly, and on a more careful view it turns out to be really something totally different in all its parts. We see partly with our judgment, and partly with our eyes, and it often happens that our judgments warp and change the impressions on the eye. The discoverer of the change of Sirius from its former white, had been so long accustomed to regard that star as of a purple or violet blue, that it was some time after I had said it was green, before he convinced himself of its green color. Hundreds of observers had seen Sirius through a telescope, and yet Clark, of Boston, was the first to notice that it had a companion, although that companion had been plainly enough in the field of view of all their telescopes. Previous observers did not see it, although they must have received the impressions on their retinas.

In a clear night we seem to see, by an optical illusion, ten thousand stars. The whole heavens swarm with them, and all, on account of their minuteness, appear to the naked eye to shine with a white light. The milky way deepens this general impression of whiteness. Probably less than fifty stars on any night, at once, are large enough to give the impression of colors to the naked eye. Thus the great mass appearing white, we assume that all are white, and by this means, the judgment being wrong, the colors strike the retina, but are not noticed.

While the telescope is necessary to distinguish the colors of the smaller stars, I have a suspicion that the naked eye is best for stars of the first magnitude, and perhaps for the second. These can be seen sufficiently well by the unaided eye, and no delicate tints are absorbed or added, as may possibly be done by the glasses and specula of instruments. The disturbing effects of the atmosphere, of moonlight, and of artificial lights, may be avoided by repeating the observations at different times.

Catalogue continued, November 10, 1863.—Altair and Deneb, or Alpha Cygni. The former of these stars was described by Humboldt in 1850 as yellow and the latter as white. They were numbered 18 and 19 in this Catalogue, and announced to be blue in June, 1863. I had watched them for several months nearly every clear night, and on the 20th of August I first noticed that they were green. On all good nights since then they have appeared to myself and to others, whose opinions I have solicited, to be conspicuously green; but on damp, slightly hazy nights, from the effects of the atmosphere, they appear blue. It is rather oppressive for me to make this announcement, for I have been obliged already, in a former communication, to say that two other large stars, Sirius and Vega, had changed from blue to green. Certainly this change does not arise from any peculiarities of vision, for I have in all cases carefully consulted the views of others and found them to accord with my own. I know not how to attribute the change to atmospheric causes, for I had observed them all, except Sirius, to be blue several months before, in all weathers.

20. Arcturus.—This is one of the stars denominated red by the ancients. In modern times, according to reliable observations, it has changed its color. J. F. Julius Schmidt, formerly of Olmutz, recently made Director of the Astronomical Observatory at Athens, and distinguished for his observations on variable stars, which he communicated to the Ast. Nach., says, that for eleven years he had considered Arcturus to be one of the reddest of the stars, and, especially in 1841, he had ranked it in color with Mars. To his surprise in 1852 he saw it to be yellow, and entirely destitute of any reddish hue. It then appeared to him by the naked eye lighter than Capella. Capella two years before had been described by Humboldt as yellow, with scarcely a tinge of red; since then Capella has become blue. During the present year (1863), I have dozens of times and in all weathers, observed Arcturus to be decidedly orange, and of a clear, beautiful color. In this I have been confirmed by other observers. The colors of Arcturus may therefore be stated as having been red, yellow, and orange.

References have already been made in this Catalogue to the changes of color in double and multiple stars. The numbers, such as $3 : 7\frac{1}{2}$, immediately after the names of the following double stars, indicate the magnitudes of the companions. The authorities are given after the colors. Some of these I have taken from the original papers, and some I have not so verified, but presume them all to be correct:

21. 95 Herculis, 5 : 5.—Hitherto catalogued as a di-

versely colored pair of stars to an extreme degree: one being described as apple green and the other as cherry red, and also as an astonishing yellow green and an egregious red. In 1856–'58 they were nearly colorless and without any diversity of tint, and in this latter manner they were described by Struve in 1832–'33, and by Sestini in 1844–'45. Hence a probability of their being colorless once in about twelve years.—*C. Piazzi Smyth.**

In the November number, 1863, of the "Monthly Notices of the Royal Astronomical Society," a suggestion is made, from very high authority, Professor Airy, that because the changes in the two companions have in all these cases been simultaneous, they are liable to the suspicion of having been produced by instrumental causes. But this apparent simultaneousness of change in both stars may have been produced by a real change in only one of them. If the two stars were white and one of them were to change to an "egregious red," then by contrast in close proximity, from the well-known principle of complementary colors, the other would necessarily appear green. The operation of this principle has been very conspicuous in this city during political demonstrations and celebrations, when bright red lights have been kept burning in the streets. The ordinary gaslights all around them have appeared strongly green. It is submitted that this cause for the simultaneous change in both stars, is more probable by far than that three different instruments, in the hands of three different men, in three different countries and at as many different periods, should all, from some unknown cause, fall into the same error; and this not when directed at the stars generally, but only when pointed to a particular one.

---

* See the Proceedings of the British Scientific Association for 1863.

CATALOGUE OF DOUBLE STARS.

22. Mizar, Zeta Ursæ Majoris, 3 : 4.—Both greenish white. Struve.
   White and pale green. Webb.
   Both yellow, the 4 has the deeper hue. Mitchell, 1860, April 30.*

23. Xi Bootis, 3½ : 6½.—Orange and purple. Webb.†
   Pale yellow and orange. Mitchell, 1862, July 6.

24. 32 Eridani, 5 : 7.—Bright yellow and flushed blue. Webb.
   Orange yellow and pale blue. Mitchell, 1861, Jan. 31.
   Yellow and pale green, very decided. Mitchell, 1862, Dec. 28.
   Yellow and green. Mitchell, 1863, Jan. 1.

25. Gamma Virginis, 4 : 4.—Silvery white and pale yellow. Webb.
   Both yellow. Mitchell, 1860, Feb. 20.

26. 35 Piscium, 6 : 8.—White and purplish. Webb.
   The 6 is light yellow. The 8 is peculiar; there is a brown mingling with its reddish light. Mitchell, 1860, Jan. 2.

27. 23 Orionis, 5 : 7.—Greenish white and white. Struve.
   Creamy white and blue. Webb.
   The 7 is of a darkish color. Mitchell, 1860, Mar. 6.

28. 39 Ophiuchi, 5½ : 7½.—Pale orange and blue. 1838.
   The 7½ yellow. Sestini, 1846.
   "   bluish. Smyth, 1851.
   "   clear blue. Webb, 1854.

* See "American Journal of Science and Art," July, 1863, for Miss Mitchell's observations.

† For several valuable popular papers on the double stars, by the Rev. Mr. Webb, see the first four volumes of the "Intellectual Observer," London.

7

29. Polaris, Alpha Ursæ Minoris, 2½ : 9½.—Yellow and d
white. Struve.
Yellow and blue. Sestini, Dawes, Webb.

30. Iota Cancri, 5½ : 8.—The 8 deep garnet, Feb. 8, 178
bluish, Dec. 28, 1782; and blue, Mar. 1
1785. Herschel, Sr.
Pale orange and clear blue. Webb.

31. Sigma Scorpii, 4 : 9½.—The 9½ white. Sestini.
Dusky and plum color. Webb.

32. Delta Corvi, 3 : 8½.—The 8½ white. Sestini.
Pale yellow and purple. Webb.

33. Pi Bootis, 3½ : 6½.—Both white; a ruddy tinge son
times in 6. Webb.

34. Alpha Herculis, 3½ : 5½.—" Intense cærulea." Stru
Orange and emerald. Webb.

35. Delta Serpentis, 3 : 5.—Yellow tints. Dembowski.
Bright white and bluish white. Webb.
Both bluish. Webb.

36. Eta Cassiopeæ, 4 : 7½.—" Flava et purpurea." Stru
Fletcher.
Red and green. Herschel, Jr., South.
Yellow and orange. Sestini.
Dull white and lilac. Webb.

37. Iota Bootis, 4½ : 8.—The 8 azure. Sestini.
The 8 lilac. Webb, 1850.
Light yellow and dusky white. Webb, 1850.

38. 39 Bootis, 5½ : 6½.—White and lilac. Some writ
ascribe a bluish and some a ruddy tint to t
Webb.

39. Epsilon Lyræ.—The two companions of this double star are designated Epsilon 1 and Epsilon 2. Each of these again is double.
Epsilon 1, 5 : 6½. Yellow and ruddy. Webb.
During five years the 5 was bluish. Struve, Dembowski.

40. Gamma Cygni, 4 : 7.—Both white. Herschel, Sr.
Viridi-cærulea. Struve.
The 7 light emerald. Smyth, 1839.
Golden yellow and flushed gray. Webb, 1850.

41. Beta Lyræ.—A quadruple star; 3 : 8 : 8½ : 9.

42. Gamma Lyræ, of third magnitude.—Both these stars, Beta Lyræ and Gamma Lyræ, seem to be changing their colors. Herschel, Sr., and South gave Beta as white. Next, Smyth, in 1834, gave the general impression as white, the four companions being in the following order: very white and splendid, pale gray, faint yellow, light blue. He gave Gamma Lyræ then as being bright yellow. Schmidt regarded the colors of both Beta and Gamma the same—yellowish white—from 1844 to 1855. Webb, in 1849–'50, regarded Gamma as much less yellow than Beta, if not white. In 1862, the latter observer found Gamma the paler in tint, though the difference was not considerable. According to these statements, Beta changed from white to yellow and Gamma from yellow to white. Both were of the same color—yellowish white—according to Schmidt, about 1844. The only discrepancy is Schmidt, for the latter portion of this time, the former portion being remarkably confirmative.

43. Eta Lyræ, 5 : 9.—Cærulea. Struve, during five years, about 1830.
Sky blue and violet. Webb, 1834.
The 5, yellow. " 1849–'50.
" pale yellow. " 1862.

44. Gamma Andromedæ, 3½ : 5¼.—Deep yellow and sea-green. Webb, 1862. The 5¼ is double, and the colors of the two latter have been given as follows:
Subviridis et violacea. Secchi, 1856.
Yellow and blue. Sir W. K. Murray, 1857.
" " Dawes, Jacob.

45. Gamma Arietis, 4½ : 5.—Both "egregie albæ." Struve, 1830.
White. Dembowski, 1852, 1854, 1856.
The same, either white or light yellow. Piazzi Smyth, 1856.
Full white and faint blue. Webb, 1862.

46. Iota Trianguli, 5½ : 7.—White or yellow and blue. Secchi.
Topaz yellow and green. Webb, 1862.

47. Gamma Ceti, 3 : 7.—The 7 tawny. Webb, 1850.
Pale yellow and lucid blue. Webb, 1863.

48. Gamma Leonis, 2 : 4.—White and reddish white. Herschel, Sr.
Bright orange and greenish yellow. Webb.

49. 72 P. II. Cassiopeæ, 4½ : 7 : 9.—White, blue, ruddy violet. Dembowski. 1854–'56.
Pale yellow, lilac, blue. Webb, 1863.

50. Kappa Cephei, 4½ : 8½.—The 4½ greenish. Struve.
Pale yellow and blue. Webb, 1863.

51. Zeta Cephei, 5 : 7.—Yellowish and blue. Struve, 1831.
Both bluish. Smyth, 1839.
White and tawny or ruddy. Webb, 1850.
Flushed white and pale lilac. " 1851.

52. 40 Draconis, 5½ : 6.—Both white. Struve, 1832.
Both white. Webb, 1839.
Both white or yellowish. Webb, 1850.
Both yellow, the 5½ deeper. " 1856 and 1863.

53. 12 Canum Venaticorum, 2½ : 6½.—White and red. Herschel, Sr.
" With all attention I could perceive no contrast of colors in the two stars." Herschel, Jr., 1830.
Both white. Struve, 1830.
Yellow and blue. Sestini, 1844.
Full white and very pale white. Smyth, 1850.
White or a little yellowish, and tawny or lilac. Webb, 1850.
Pale reddish white and lilac. Smyth, 1855.
White and pale olive blue. Dembowski, 1856.
Same as in 1850, but with very little contrast. Webb, 1862.
Flushed white and pale lilac. Webb, 1862.

54. Sigma Coronæ, 6 : 6½.—Creamy white and smalt blue. Webb, 1862.
The 6½ has had many changes as follows : certainly not blue and differing very little from the other. South, 1825.
White. Struve, 1836.
A yellow ashy and doubtful blue. Dembowski, 1854–'57.
Sometimes blue, sometimes yellow. Secchi, 1855–'57.

"At one time ruddy, at another time bluish, apparently changing white being looked at; a versatility of hue which I have remarked in other stars similarly circumstanced." Webb, 1850–'55.

55. Mu Cygni, 5 : 6.—White and pale blue. Struve, 1831.
Yellow and more yellow. Sestini, 1844.
Reddish yellow and olive. Dembowski, 1853–'54.
Clear light yellow and ashy yellow. Dembowski, 1855.
"The 5 yellow, while the 6 showed the curious effect of an undecided and changeable hue—blue and tawny." Webb, 1850–'51.
The 5 yellow. Webb, 1862.
"Secchi's colors are here uncertain and variable."

56. Alpha Piscium, 5 : 6.—Greenish and pale blue. "There seems to be something peculiar in the color of the smaller star, as to which observers are strongly at variance with each other, and even with themselves. Some see no contrast, some agree with Smyth, some find it tawny and ruddy. The details are curious, but too long for insertion here. Other small stars show a similar uncertainty." Webb.

The frequent changes in some of these stars—the last three or four of this Catalogue especially—are remarkable, and seem inexplicable to astronomers. I presume the difficulty arises, not as is supposed from the atmosphere, or from the instruments, or from personal peculiarities, but chiefly from the frequency of the real changes in the stars. If, for instance, it be complained that "Secchi's colors are uncertain and variable," it is because in such instances the colors of the stars are uncertain and variable.

A few remarks should here be made to remove the objection that possibly some of the apparent changes of color of the stars may be merely the errors of observation, or the effects of the atmosphere, and not real changes in the celestial bodies. Nevertheless, an apparent change is a fact in the constitution of the world, and deserves a notice and an explanation. From whatever causes these changes may arise, there is needed a faithful collection of all the facts in this department of astronomy. They are scattered about in many volumes and many various scientific depositories, and no one, as far as I am aware, has brought them together or made them a special study. But in making such a collection, or catalogue, no changes should be omitted. Whether we regard them as apparent or real, whether they be small or great, whether they may have been slow or sudden, none should be suppressed by the compiler in his catalogue. To admit some and reject others because in his opinion some are right and others are wrong, would be making his work a confused medley of facts and opinions unworthy of reliance. If, as appears undeniable, there be changes in the colors of the stars, then, from the nature of things, there may be small changes as well as great ones. To reject a recorded change simply because it is small, would therefore be a real misrepresentation of the case, and a virtual falsification of the records. Moreover, the colors as they stand recorded are from experienced observers—men whose lives have been devoted to an accurate representation of facts, who do nothing without care and deliberation, and whose common and avowed practice is not to record any color when the atmosphere is not favorable for such observations.

As already stated, there are difficulties in deciding on colors by the naked eye when the star is not large, and when the departure from white is small. But this difficulty is not in the way of large stars, as Arcturus and Sirius; nor does it apply to the telescope, except in the very smallest magnitudes. To decide between two different colors, such as red and blue, is never difficult; and when two colors are blended, it is the custom to name them both, as bluish green, reddish yellow, and the like. The disturbing effects of the atmosphere, or of the instrument, may be detected either immediately or after several nights of observation. The atmosphere cannot color one star and leave all the other stars in the same neighborhood uncolored. The telescope cannot act peculiarly on any one star; it must treat all alike, especially of the same magnitude and color. Simple comparison is therefore an admirable test; and another important test is time—watchfulness every night through different changes of weather. If hereafter even this shall not be found satisfactory in any one locality, then simultaneous observations at widely distant places will most certainly eliminate all suspicion of mistake. For instance, observations may be made at Australia, the Cape of Good Hope, and Chili in the southern hemisphere; and in the northern hemisphere both on the Pacific and Atlantic coasts of America, on the Atlantic coast of Europe, in Russia, and in Hindostan. If the star shall prove of the same color at all these different regions at the same time in favorable weather, then that color may be regarded as unquestionable. Even by using one locality alone absolute certainty may be acquired —as the red colors of Aldebaran, Betelguese, and Antares. The same certainty may be looked for in this as in other departments of astronomy, and even greater certainty than in many. There is an uncertainty, in the opinions of wise men, of three millions of miles in the distance of the earth from the sun; and yet this uncertain distance is used as a measuring line to fathom other and far greater distances. But this uncertainty to so large a degree does not take away from the supreme value of the determinations nevertheless. These determinations, with all their known reservations, are held as of the highest importance. So in the colors of the stars; mistakes may be made, the intermixtures of error may certainly exist, though we cannot tell exactly where they are, and yet the present recorded observations are precious beyond estimation. And a time is coming when simultaneous observations from various positions in both hemispheres will render them beyond suspicion. To hasten on this time we have only to make good use of the materials already on hand.

Why the changes in the colors of the stars are not more frequently observed, was pointed out in a former communication. Why the belief in their real occurrence is hard to be admitted, and why their observed changes are ascribed to supposable errors from the instruments, from the atmosphere, and from personal deficiencies, seems to arise from the opinion that such vast bodies cannot possibly undergo great changes in a short time. But this opinion rests on no known scientific grounds. When fairly viewed, the fixedness of the colors of the stars should not seem more likely than the fixedness of their positions. Indeed, the two ideas are very much alike. In ancient phrase, the stars were said to be "riveted" to the vault of heaven. Now we know from observations more refined that many of them move, and we have a conviction, from the nature of attraction, that they must all move. In like manner, in a universe where every known object is subject to change in various ways, our first ideas should be that the colors of all the stars must change. Hence we should approach the recorded changes with favorable judgments. If we are to have any prepossessions in the case, they should be that the changes are real in the stars themselves. And when we reflect on the habitual caution of long-experienced observers, men whose very existence is devoted to the accurate delineation of facts, we should place a high reliance on their recorded observations, and not think that they have lightly allowed themselves to be imposed upon by optical illusions.

I cannot hope to be able to add any thing to the knowledge of practised observers respecting the sources of error and the rules to be observed in making observations; but as these have never, that I am aware, been embodied in print, I offer the following, chiefly for the assistance of the many who may be disposed to observe the larger stars with the naked eyes. Such stars are indeed very few, but the observations may be the more useful from being made frequently and by many persons:

1. Damp and slightly hazy atmospheres make a green star appear blue.

2. Moonlight greatly obscures the colors of the stars, giving them a yellowish hue. Twilight has a like effect.

3. Rising up before daybreak in the morning, or at any time after a couple of hours of sleep, totally disqualifies my eyes for observations on the colors of the stars. They then all appear whitish. I do not know whether my own eyes are peculiar in this respect, but I think they are not.

4. Artificial lights reaching the eye obscure the colors of the stars.

5. On account of the faintness of the light of the stars, the eye often requires to be fixed upon them for a considerable time before their impressions take full effect.

6. Comparisons between neighboring stars, and some practice in star observations, are often necessary to decide on the real colors of the stars.

7. The atmosphere must have like effects upon similar stars in the same neighborhood. Hence a peculiarity observed in any star may be brought to a determination.

8. Observations on the same star during a considerable

interval of time and through different changes of weather, may aid in giving confidence to a determination.

9. Perfect independence and candor are necessary. Our previous judgments are apt to warp these delicate impressions on the retina, and whether we have derived these judgments from ourselves or others, we must be careful to lay them completely aside. For want of doing this we may not notice a change of color, although such a change may be before our vision.

10. Personal peculiarities of vision may be ascertained by consultation with others.

11. Discrepancies between the accounts of two observers may arise from differences of dates; hence, in apprehension of sudden and frequent changes in the stars, the dates of observations should be carefully given.

THE INFLUENCE OF THE EARTH'S ATMOSPHERE ON THE COLORS OF THE STARS. JUNE, 1864.

From the small amount of attention paid to the colors of the stars as a distinct branch of physical research, a vague and indefinite impression has been somewhat prevalent that the atmosphere of our earth has great power in producing the apparent colors and the changes of colors of the fixed stars. The subject is highly important. During the last two or three years it has occupied much of my attention, and I propose in this paper to present my method of investigation and the results to which I have been led. To ascertain what the influence of the atmosphere might be, I selected for special observation a few of the larger stars, taking some of the red, some of the blue, some of the green, some of the yellow, and some of the white. So many different classes of stars watched carefully during the various changing conditions of the atmosphere, seemed most likely to yield valuable conclusions.

1. The red stars were Aldebaran, Antares, and Betelguese. These are all of different shades and intensities of red. In proportion as the atmosphere loses its transparency by the condensation of moisture, these stars lose their distinctive peculiarities. Their redness gradually becomes obscured, and they at last appear of a dull, unsatisfactory white.

2. The blue stars were Capella, Rigel, Bellatrix, Procyon, and Spica. Some of these, as first Procyon, and then Rigel, are far more intensely blue than the others. But as the atmosphere becomes thick and more impervious to distinct vision, their different intensities of blue fade away, and the observer is at length puzzled to decide of what color these stars really are. He feels safest in announcing that they seem white, though not of a clear, decided whiteness.

3. The green stars were Sirius, Vega, Altair, and Deneb or the largest star in the Swan. These stars were observed to be green by myself, in the following order: Sirius in the autumn of 1862, Vega in June, 1863, and Altair and Deneb in August, 1863. It is remarkable that a very slight haziness in the sky completely hides their green color, and causes them to appear unmistakably blue. A still thicker haziness has the same effect on them as it has on all the blue stars already described, gradually obscuring their blue color, and ranking them among the many hundreds of stars which the naked eye cannot decide to be colored.

4. The yellow star was Arcturus; this being the only one which appears decidedly yellow to my vision, unaided by instruments. Several others incline the naked eye to regard them as yellow, such as Polaris and the larger stars of Ursa Major and of Cassiopeia, but not sufficiently so to produce a firm belief. Arcturus, in a clear sky, has a fine

light orange yellow; but as the sky becomes less and less clear, the yellow fades away, and ultimately the color of this star turns to a dim white, and becomes undistinguishable from that of the larger stars of Ursa Major, with which, from their position, it may be handily compared.

5. The white stars were Regulus, Denebola, Fomalhaut, Polaris, the constellation of the Waggon, and several others of the second and third magnitudes. They may be called white stars with reference to their appearance to the naked eye, to mine at least, but we are not bound on that account to believe them to be really white. As they are not first magnitude stars, they probably seem white to the unaided eye only because their light is not sufficiently great in amount, or intense in color, to appear colored. There may be persons with unaided vision acute enough to perceive their true colors. But whatever may have been the conditions of the atmosphere, I have never observed them to be other than white. No changes of the air have had the power of presenting them in any shade as colored stars.

Thus the influence of the atmosphere of our earth upon the stars of all the different colors, according to these observations, is the same. Whether the stars be red, blue, green, or yellow, the effect of changes in the atmosphere is to rob them of their peculiar shades and intensities, and to reduce them all to a dull, colorless condition—a dim whiteness, in which their indistinctness produces a feeling of uncertainty and doubt in the beholder. Nor in any case have I seen any change in the atmosphere turn a star from one color to another, except from green to blue, and this is simply reducing one shade to another; for green, like purple, is but one of the modifications of blue. I have never seen a red star become blue, nor a blue star become yellow, nor any other similar change by any change in the atmosphere. If such an occurrence were possible, I believe

I would have observed it during the past two or three years. The effect of moonlight in obscuring the colors of the stars, and giving them a yellowish shade, can hardly be called an atmospheric action. Neither can the effect of the rays of the sun in the earliest daybreak of the morning or in the latest twilight of the evening, be called an atmospheric operation. Such an effect tends to impart a general whiteness to the stars, obliterating their colors in part or in whole, the same as in the end it obliterates all their light.

The question now arises, How is it that the atmosphere, when hazy and imperfectly transparent, has the power of depriving the stars of their colors, whatever their colors may be, and reducing them all alike to a dull whiteness? The reason may be seen in the simple fact of the obstruction of their light. Their light becomes diminished in amount to such a degree that it no longer has the power to produce the sensation of color on the retina. Nearly all the stars, when viewed through a telescope, are colored; they are of some hues other than white. Of this I adduced evidences in my communication for these Proceedings in June, 1863. They appear colored through the telescope because their light is collected by the instrument in a comparatively large mass; so large that it can make their colors readily perceived. Take away the instrument from all except the larger stars, and the pencil of light becomes so small as to be without the power of imparting the sensation of color. In the same manner the pencil of light from the larger stars may be reduced by haziness in the atmosphere to so small an amount as to be incapable of imparting the sensation of color, except a dull whiteness, whatever their real colors may be.

But how does it happen that a green star is changed by haziness to blue? I once thought that possibly this effect might be due to the same cause which makes the deep ocean, the distant mountains, and even the atmosphere, appear blue. After further observation and reflection I cannot adopt that explanation; for then all the stars, like the distant mountains, would be colored blue. Then there would be no such contrasts of all colors among the stars as we now behold. The true explanation seems to be that the mists of the atmosphere, in acting on the light of a green star, first obstruct the yellow rays, and after these are all absorbed then the blue rays alone will be visible, and the star must appear blue. Ultimately the mist may become so impervious that the attenuated ray of light can no more excite the sensation of color, and the star must appear dimly white.

Before it can be admitted as a scientific truth that the atmosphere of our earth has the power of changing the color of a fixed star from one hue of the rainbow to another totally different, there must be brought forward a number of well-authenticated facts as grounds for such a belief. We must have the specifications of certain stars which have been seen to change, and the dates of such changes, and the conditions of the atmosphere by which such changes have been produced, and also the numbers and the names of the persons by whom such phenomena have been witnessed. . Such evidences of the changes of the colors of the fixed stars by our atmosphere have never been seen nor heard, and for my part, judging by my own observations, I never expect to see them, nor to hear of them. An exception to this remark may be the case of a green star turning to blue, as already explained. Perhaps another exception may yet be found, as indicated in the following passage from Humboldt. The italics are not in the original: "We do not here allude to the change of color which accompanies scintillation, even in the whitest stars, and *still less to the transient and generally red color exhibited by stellar light near the horizon*, a phenomenon owing to the character of the atmospheric medium through which we see it." The turn of the expression "still less" shows that he regarded the matter as inconspicuous and unimportant, and the remark is made only in a casual manner. Nevertheless, incidental as the remark may seem, it is the most precise and circumstantial I have found in any author on the influence of the atmosphere on the colors of the stars. But is it really true that the atmosphere can impart a transient and generally red color to stellar light when near the horizon? In the absence of all confirmation to the above remark of the distinguished philosopher, I selected as test stars Vega and Capella, both first-magnitude stars, the former green and the latter blue, and the one or the other is grazing the northern horizon nearly all the year. But I have been unable to detect the changes he mentions. May not his remark have arisen from observations on the planetary bodies, and have been inadvertently extended to the fixed stars? The planets, especially Jupiter and Venus, according to my observation, are sometimes, though rarely, reddened like the sun and moon by the atmosphere. But whether Humboldt's assertion be confirmed or not, it cannot effect our decision about the real changes of the colors of the stars. No one would pretend to announce a change in the color of a star simply because of a "transient" appearance of a change while near the horizon. In the same manner, probably, the idea has got afloat unguardedly that, because the atmosphere of our earth has the capability of giving occasionally a red color to the sun, moon, and the planets, it must therefore have not only the same effect on the fixed stars, but even the power to turn them to all the hues of the spectrum between red and blue. But this rapid generalization is no more warranted by sound reasoning than by observation. The sun, the moon, and the planets, have sensible disks, which the fixed stars have not. Hence the optical phenomena of these two classes of bodies, differ widely. The fixed stars, under the influence of our atmosphere, are made to scintillate; they then twinkle with an unsteady light, and to good eyes they flash out rapidly and fitfully all the varieties of colors. This shows the difference, in an optical point of view, between the fixed stars and the other celestial bodies, and the impropriety of a hasty generalization from one class to the other. Because the atmosphere can redden one class it by no means follows that it can redden the other, much less that it can impart to the other all imaginable hues.

Another cause for the belief that the atmosphere can impart different colors to the stars, may be found in the necessity for some explanation of their changes of color. It is assumed, though without any known reason, that the intrinsic colors of the stars cannot change, at least in the space of two thousand years, and hence there is a necessity for an explanation of their apparent changes in some other way; and as the handiest method these changes are attributed to the atmosphere of our earth. That the various colors of the stars are not produced by our atmosphere, nor by optical instruments, nor by personal peculiarities of vision, becomes perfectly evident from the following simple consideration: If their colors were produced by any one of these causes, then there would not be that beautiful contrast of colors which we now behold; then it could never have been said of the cluster Kappa Crucis, that the various bright contrasted colors of its different members give it all " the effect of a superb piece of fancy jewelry." Instead of this there would be in that cluster, and in every other region, a dull, monotonous color in all the stars alike. It has happened that travellers, in coming from Europe to America, have expressed their surprise at the beauty of our sky, when noticing for the first time in their lives the different colors of the stars. This has been supposed to be the work of our atmosphere, the natural operation of the gaseous envelope of our earth. The true explanation is this: The stars appear colored to the naked eye in Europe as well as in America. Astronomical observers see them colored the same in all countries. But in some countries their colors are slightly dimmed by the more habitual haziness of the atmosphere, so much dimmed that they are not noticed by unprofessional gazers. When these latter persons arrive in a more cloudless region, they notice the colors of the stars simply because a slight veil is withdrawn and not because new colors have been added.

The evidences of changes of color are now most abundant among the double and multiple stars. This is because the colors of these have been more generally recorded. Hence the importance of having records made, as frequently as possible, of the colors of all the stars, as they appear both to the aided and the unaided vision. The elder Struve compared his own observations on the colors of the double and multiple stars with those of the elder Herschel on the same stars. The differences of color in many, after a long interval, were wonderful. See particularly Struve, third Catalogue, published in 1837, entitled "Stellarum Compositarum Mensuræ Micrometricæ." Here end the extracts from my papers on the colors of the stars, in the proceedings described on page 132.

THE CAUSE OF THESE CHANGES OF COLOR.

The explanation I have already offered for these changes of color seems open to no objection that I can perceive. The different stars consist of very different mineral elements, as the Fraunhoffer lines plainly show. Therefore, they should burn with differently colored lights. In the same star sometimes one set of elements may have a predominance in burning at the surface, and at other times another set, and therefore it should change its color. This alternation among different sets of elements in burning at the surface, results from the fact that some elements are lighter and more combustible than others. The lighter and more combustible must burn first at the surface, and when they expire, then the more dense and less combustible may take their turn. Our earth may serve as an example. We cannot put down our feet on the ground or on the rocks, without treading on phosphorus, sulphur, and potassium. They are in our food, and they form part of our bodies. Hence the plants must get them everywhere from the ground, even where our analyses cannot detect them. They must, therefore, exist in immense quantities in the crust of our earth, and for illustration we may suppose them still more abundant. Being very light and very inflammable, they would burn on the surface before other heavier and less combustible elements. Phosphorus would give a yellow light, sulphur blue, and potassium, when burned on water, gives a beautiful purple. After they had all combined, then other heavy and less combustible elements, as copper, would begin to burn, and a very differently colored light might be the consequence. Hence our earth as a star would change its color, and this might be repeated suddenly and often.

Besides the lightness and combustibility of an element, another important item would determine its priority in the succession of burning, namely, the period when the element was created. We do not know that they were all created at the same time. Probably, and we may say certainly, they were not. The chemical elements are mere modifications of matter. They have been moulded by powerful physical forces out of some previous general form; and the date and place of their creation would have much to do with the date of their burning. We do not yet know the exact history of these events, so far distant in the realms of the eternity that is past; but there is unmistakable evidence that our globe was once wrapped in a great conflagration. Nothing can be more combustible than the elements of which it is composed. We see a likelihood that some of these elements would burn before others. Hence we may use our planet, which is really a star, as an illustration of the colors and changes of colors among the other stars. In this way we see that the chemical theory clears away all the darkness which has been hanging over some of the most momentous facts in creation; and as these facts are so numerous and so diverse from one another, we must regard that theory as true, especially as there can be raised against it only one ill-founded objection. That one ill-founded objection should not overrule many dozens of facts on the opposite side, that are well founded.

SECTION IX.

REVIEW OF EVIDENCES.

In this survey of the fixed stars we have seen a large number of facts which tend to prove that their light and heat are caused by chemical action. All the curious changes in their colors and in the amounts of their light can be explained easily and naturally by this theory. We should now make a brief recapitulation of these facts, and this can best be done in drawing a comparison between the sun, the fixed stars, the earth, and the moon.

In the sun we behold the chemical elements, the best of all imaginable fuel. We behold these elements so intensely heated, that they float as vapors in the flames more than a thousand miles high. Then there is the melted body of the sun, a vast fiery ocean, with its rapid currents and violent agitations, and similar agitations in the raging flames. The clouds rising 80,000 miles above in the immense atmosphere, and the indescribable flood of light and heat radiating away, constitute the most wonderful and the most truthful of all pictures of chemical action in the form of combustion. The pores and the larger spots give the only evidences of a cooling and solidifying tendency; and the deficiency of heat in the spots, their cracking, their breaking into fragments, and their ultimate melting, are important features, as connecting the sun with our earth and the variable stars. The very great sizes of the spots, some as much as 45,000 miles broad, and close groups of spots more than 100,000 wide, add still further to the relationship between the sun, the stars, and the earth.

In the fixed stars we behold the chemical elements again as the fuel for their fires. The dark lines in their spectra bear the same evidence as in the sun, that these elements are in a state of vapor floating in the flames, and that beneath these flames there is a still more intensely heated body which must be above the point of fusion, and, therefore, a vast ocean of fire constituting the mass of each star. The variability of their light, and the perfect periodicity, even to a second in that variation, assure us of their rotations on their axes, and of their spots grown so large as at each rotation to hide a considerable portion of their light. This immobility of the spots assures us that they are continental in their characters, reaching quite around the stars, though still leaving much of each star uncovered. The irregularity of other periodic stars assures us of the changes of the spots, both in their sizes and in their positions; they can float about like floating islands, the same as the spots in the sun. And as our sun is often free from spots, so some variable stars are for a long time perfectly invariable, until, by irregularities in chemical action, a cooling tendency again allows the formation of spots. The irregular routines in the increase and decrease of light show that the spots are irregular in their contours; and the changes of these routines indicate that in the stars, as in the sun, the spots change their forms. As the maxima and the minima in the light of the periodic stars, are sometimes greater and sometimes less, we have reason to believe that in the stars, as in the sun, the sizes of the spots greatly vary. When the routines of irregularities do not change, the periods do not change; and when the routines change, the periods change. These are two remarkable facts, and they clearly indicate that the changes in the spots cause the changes at the same time both in the routines and in the periods; and that to the absence of changes in the spots are due the absence of changes at the same time both in the routines and in the periods. The increase of light in a periodic star is more rapid than its decrease, and this indicates that, from causes before explained, the smaller end of the spot precedes the larger. Stars with two maxima and two minima in the same period, indicate that they have two dark spots on different sides; and as one maximum or minimum is often larger than the other, this merely shows that one of the two spots is larger than the other. Some stars show an extreme variation in amount of light without any signs of periodicity, as for instance Eta Argus, and this is easily explained by irregular energy in chemical action. But by the immutable laws of chemical action all fires must ultimately expire for want of fuel. The greatest conflagrations must come to an end. The spots on some stars are larger than on others. Between the maxima and the minima of some stars, the difference is almost imperceptibly small; in others it is so large that at each rotation the stars at their minima are invisible. In Mira Ceti this invisibility continues more than two hundred days, which is two-thirds of its entire period of rotation. The next great consequence of expiring chemical action, is a complete growth of the spot around the star, and the entire covering up of the light and heat in the interior of the star, where they can make themselves known only by volcanic action, earthquakes, and thermal springs. It is a lost star. As in the irregularities of chemical action, even phosphorus, after burning out, may rekindle again; and as spontaneous combustion sometimes surprises the most careful chemist, so a lost star may break forth again into a temporary glare after its fires have been so nearly extinct as to be invisible at our distance. These are called temporary stars. The very different bright colors displayed by different stars, show that the chemical elements there, true to their natures, burn some with one color, and some with another; and the changes in the colors of the stars indicate that when one set of elements is consumed, other sets of elements, with lights of other colors, take their turns.

In our earth we see again the chemical elements. The way they are combined proves, beyond all doubt, that chemical action once made our planet as brilliant as any star in the heavens. If chemical action lighted up one star, we may be sure that it lights up all other stars; especially when we behold such strong resemblances in light and heat between the sun, the fixed stars, the earth, and the moon. The present crust of our globe, resting on a fiery ocean beneath, is but the spots grown large, completely investing the star, shutting out its light except in volcanic eruptions, and adding one more to the catalogue of lost stars. Before its crust had grown so large as to envelop the entire globe, we can easily understand how our earth must have been a periodic star. And remembering the different elements of which it is composed, we can plainly see how it might have changed its colors while these elements, one after another, had the predominance of burning on its surface.

The moon exhibits in high relief an advanced stage of the fiery history of our earth. There the process of cooling has gone so far that even volcanic action has ceased. But there stand the volcanic craters, there lie the cold hard lava streams; and their sharp outlines, their gigantic proportions, and their surprising numbers, leave not a doubt of the most intense igneous action. Even the once unaccountable fact that the moon's rotation on her axis, and her revolution round the earth, are performed in precisely the same time, now tells the eloquent story of her former igneous fluidity, her gradual cooling, and the slow formation of a rigid crust. That rigidity opposed the tide wave in the fiery ocean, until that opposition delayed more and more her rotation on her axis, and at length that rotation came to coincide with her revolutions round the earth.

The general history of the great globes of space, whether we call them fixed stars, suns, planets, or moons, has the following features: their changing from one bright color to another, their gradual progression from light to darkness, their slow advancement from heat to cold, and from a liquid to a solid condition. This course of changes in them all must have the same course. It is identical with the cause of changes in that form of chemical action called combustion, or burning, which, when several elements are concerned, proceeds in like manner from one color to another, from light to darkness, from heat to cold, and from a liquid to a solid condition.

Thus the parallelism between the sun, the fixed stars, the earth, and the moon, is complete when compared in the light of the chemical theory. It is equally as complete as when they are compared in the light of ordinary astronomical data, which was done in the very beginning of this volume. Both the methods of comparison demonstrate that all these great bodies have essentially the same nature and the same general history; and that their differences are mere matters of degree in size, density, motion, heat, and luminosity—some of these differences depending on their different stages in the same progress. A third method of comparison is by the nebular theory, which also gives the same origin, the same history, and the same general nature to all the stars, whether they be called suns, planets, or satellites. The fact that all these three methods of comparison bring us to the same result, confirms our confidence in that result, and in the methods of comparison. It is a new and an additional argument in favor of the theory of chemical action as the cause of the light and heat of the stars.

PART II.

THE FORCE WHICH SO GREATLY PROLONGS THE LIGHT AND HEAT OF THE SUN AND OTHER STARS.

SECTION X.

THE OBJECTION TO THE CHEMICAL THEORY OF STELLAR HEAT AND LIGHT, IS FOUNDED ON THREE GROUNDLESS ASSUMPTIONS.

We have now reviewed the chief evidences to prove that the present interior heat of the earth, its former entire fusion, and the light and heat of the sun and of the fixed stars, have been caused by chemical action. These evidences are strong, numerous, diverse in their nature, and they come from different and independent sources. What, for instance, can be more different and independent than the composition of the water in the ocean, the colors of the fixed stars, the spots in the sun, the fixed lines in the solar and stellar spectra, and a temporary star? Yet all these, and nearly a hundred other great phenomena, receive their explanation by this theory. When a key fits a lock with a hundred guards, it must be the true key; and when a theory explains completely such great, such diverse, and such a vast number of phenomena, that must be the true theory, especially when they can be explained by no other theory. This chemical theory would be universally believed were it not for a single objection, and this objection we will now endeavor to remove. The objection is drawn entirely from the heat of the sun. It says that, to produce all this heat, the sun has not a sufficient amount of fuel. This I shall maintain is an unfounded assumption. It assumes, first, that the materials of the sun, pound for pound, can give out no more heat than the materials of our earth. It assumes, secondly, that, even though the materials be the same, the *conditions* for producing heat in the vast laboratory of the sun, 880,000 miles in diameter, are no better, no more productive, than in our small furnaces. It assumes, thirdly, that no new chemical elements can be formed in the sun, that the materials of the sun cannot be decomposed, or metamorphosed, so as to be burned over again. Here we can decompose water, and burn the oxygen and hydrogen a hundred times; but we lose as much force or heat in the decomposing as we gain in the burning. But in the sun the case may be very different. There may be an almost infinite amount of physical force in the sun by which chemical compounds may be decomposed, and chemical combination again be resumed. It is not the force which we are to account for; that we know exists, and has existed millions of years, sending out light and heat all the time. Our object is to show that this force may ultimately reach us by means of heat from chemical action. The nebular theory, proven to be true in this volume, sheds light on this point. It proves that those modifications of matter, which we call simple chemical elements, have been produced during the nebular condensations, and therefore that they may now be produced yearly, in the sun, which has not yet reached the point of condensation required by the law of density in the solar system.

All these assumptions I shall show to be unfounded. And if an objection to a theory rests only on unfounded assumptions, that objection is worthless, especially when the theory is supported by nearly a hundred positive arguments in its favor. We will now attend to these three assumptions in the above order.

SECTION XI.

THE FIRST ASSUMPTION: THAT THE MATERIALS OF THE SUN GIVE OUT NO MORE HEAT, POUND FOR POUND, THAN THE MATERIALS OF THE EARTH.

WE have the best reasons to believe, from the nature of matter, that the materials in the sun may possibly give out more heat than those in our earth. Here in our earth one substance gives out more heat than another of equal weight. A pound of hydrogen produces more than four times more heat than a pound of carbon. Between other elements there are similar differences. It is impossible to guess how much heat may be produced by the peculiar elements of the sun; they may possibly differ from ours in heat-producing power as much as ours differ among one another in density, and this is as 256,700 to 1. The terrestrial elements differ widely among themselves on other points respecting heat, besides heat-producing power. A pound of water, to increase its temperature one degree, requires thirty times more heat than a pound of mercury. Silver conducts heat fifty times faster than bismuth. Mercury melts at forty degrees below zero, and platina at more than 30,000 degrees above. Oxygen, hydrogen, and nitrogen, are about thirty-nine times more diathermous than chlorine, and 1,195 times more than the gas ammonia. When exposed to a surface heated to two hundred and twelve degrees rock salt is ninety-two times more diathermous than Iceland spar and some other crystals. Such differences between substances in relation to heat in every possible way here on our earth, show how utterly unfounded are all assumptions about the heat-giving properties of the peculiar and unknown elements in the sun. And yet on such assumptions rests the objection against chemical action in the form of combustion in the sun.

The boundless differences in matter, the almost infinite diversities in every property between the so called simple elements, should dispose us to anticipate almost infinite diversities in their heat-giving powers in very different and distant globes. Nothing can be more wonderful than the difference between the ductility of iron and the weakness of zinc, the malleability of gold and the brittleness of bismuth, the elasticity of steel and the plasticity of potassium, the density of platina and the lightness of sodium, and still more the lightness of hydrogen. Our lives are supported in oxygen, but we die in nitrogen. Chemical diversities seem endless in number and immeasurable in extent. When from our earth we turn to the heavenly bodies, we are still further impressed with the infinite diversities of matter. Every star, as far as yet known, has a different set of fixed lines, although there are certain resemblances between them. They lead to the conclusion that each star has, in part at least, its peculiar modifications of matter, called simple elements; but the number of stars is infinite, and therefore the number of elements must be infinite. The densities of the stars lead to the same conclusion as their fixed lines. In our solar system Mercury is sixty or eighty times more dense than one of the satellites of Jupiter, and probably in a much greater proportion denser than the satellites of Saturn. This indicates a wide difference between the nature of their elements. Among the fixed stars the differences appear far wider. When two heavenly bodies revolve around their common centre of gravity, their velocities depend on their mutual gravitation. Among the double and multiple stars some move rapidly, and others slowly, and still others so slowly that their movements are as yet imperceptible. A remarkable one of the latter class is the star called Theta Orionis. It consists of four larger stars and two others extremely small, and the four larger are so situated as to be called the Trapezium. They have long been regarded as a striking object, and there they seem to stand immovable. Doubtless they move—this is rendered necessary by the universal prevalence of gravitation—but they are so excessively rare that their gravitation amounts to almost nothing, and hence their motions are imperceptibly slow. This is the case, indeed, with nine-tenths of the double and multiple stars; out of nearly seven thousand already known, the motions of less than seven hundred have been hitherto detected. In our solar system we would not naturally look for wide differences between the several members, because they are so nearly related to each other as a distinct family; and yet some are probably one hundred times more dense than others. But among the fixed stars, so distant from one another, and so slightly related, the differences in density, judging from their movements and their want of perceptible movements, must be somewhat like the difference in density between the extremes of our elements here, as 256,700 to 1.

Thus when we regard the elements of our earth, one of their striking features is their great diversities in all their properties. Among the members of the solar system these diversities are equally striking, judging from their densities. Among the fixed stars scattered far away in every direction, the diversities in their elements, judging from their densities and from their fixed lines, must be inconceivably greater. In our earth the diversities among the elements appear in their relations to heat in every imaginable respect, and among others in their heat-giving powers. And all this being the case, how unphilosophical is it in us to pretend to tell what are the heat-giving powers of the unknown elements in the sun, a body so distant and so different from our earth! The very first thought should suggest wide differences, and then the solid evidences of long-continued combustion in the sun, with such great evolutions of heat, should tell us that those differences are almost infinite.

When the Fraunhoffer lines were first interpreted by Kirchoff, there was a general acquiescence in his announcement that a few of the elements of the sun and of the earth were identical. But later discoveries have taken away the foundations of this opinion. (a) There are several thousands of fixed lines in the sun, and more are being discovered by improved apparatus, and they are so closely crowded together that it seems impossible to identify those of our earth. (b) By experiments with our elements it is found that there are perfect coincidences in the lines of several different elements, and when the coincidences are not perfect their positions are so very near as to be scarcely distinguishable. (c) In the same flame one element may cause the lines of another element to vanish; thus the blue line of strontium vanishes when the chloride of copper, sal ammoniac, or chloride of strontium are introduced in the flame. (d) A metal and its compounds give different lines, and of the same metal one compound gives different lines from another; thus the spectra of protochloride and subchloride of copper differ from each other. (e) Even with the same element in the flame a change of temperature causes a change in the fixed lines. For these reasons the attempt to identify any of our elements in the sun seems hopeless. In several stars the sodium line D has been confidently recognized, but this is annulled by a recent announcement of Secchi from Rome. "Having examined the spectra of different metals in Rumkoff's apparatus, he noticed that in the neighborhood of the sodium line, and at such small distances as could not be measured, a line was also given by many other metals, as iron, platinum, copper, zinc," &c. An American observer, Rutherfurd, after a careful study of this subject, announces that "these investigations are yet in the cradle;" and he further says, "it is not my intention to hazard any conjectures based on the foregoing observations; a great accumulation of accurate data should be obtained before making the daring attempt to proclaim any of the constituent elements of the stars."* Thus there are no facts which bear against our conclusion, that the elements of the sun are profoundly different from those of our earth. Even if a few were the same, hundreds and perhaps thousands are still different; and being different, perhaps extremely different beyond any apprehensions we now have, they may give out almost an infinitely greater amount of heat. Therefore, the first assumption is unwarrantable.

SECTION XII.

THE SECOND ASSUMPTION: THAT IN THE VAST LABORATORY OF THE SUN, 880,000 MILES IN DIAMETER, THE CONDITIONS FOR PRODUCING HEAT ARE NOT DIFFERENT FROM THOSE IN OUR SMALL FURNACES.

This assumption is unwarrantable, for the following reasons: Heat is a mode of motion. In the combination of even the same elements more violent motions, and consequently more intense heat, may possibly be produced by different conditions. Some of these different conditions in the sun will here be named:

(*a*) Combustion with us is always between a gaseous body and another which may be either gaseous, liquid, or

---

* "American Journal of Sciences" for January and July, 1863.

solid. In the sun the chief combustion takes place between liquid materials, for the liquid body of the sun is the hottest.

(b) Pressure exercises an important influence on combustion. A candle flame, burning on a mountain top and down at its foot, exhibits very different phenomena. A mixture of oxygen and hydrogen when cool, does not unite, but if a smooth platina plate be introduced, the pressure of adhesion unites the gases. The science of chemistry affords many instances of the connection between pressure and chemical action. The pressure exercised by the atmosphere of the sun must be inconceivably great, considering the height of that atmosphere, and the powerful attraction of the sun; but even this is as nothing compared with the pressure in the liquid body of the sun many thousand miles down. Think of the vertical currents in the sun, and the change in pressure from the surface to the regions near the centre! Our most powerful presses and trip-hammers bear no comparison.

(c) Heat exercises a wonderful influence on chemical affinity, and to promote rapid combustion and a more intense heat in an iron furnace, the hot blast is applied. In the sun all is hot blast.

(d) From the most recent researches heat is believed to arise from the rushing together of the atoms of two bodies in the process of combination. The collision of those atoms produces the vibrations which we call heat. Whatever adds force to those atoms in that collision, must add to the heat. The high temperature and the pressure in the sun must have an important influence on chemical combination there, and hence perhaps in adding force to the collision of the combining atoms.

(e) The force of chemical attraction, the medium, whatever it be, which impels the atoms to unite in a collision, may be more dense, or in some way more powerful, in the sun than here on our earth. This idea may be illustrated by the analogy of our atmosphere. Around the moon there is apparently no atmosphere; around the earth there is one of moderate height; around the sun there is one of enormous proportions. It may possibly be that the medium in the sun, causing chemical union, is more abundantly and more intensely aggregated like the gaseous atmosphere, or like the force of gravitation. Hence there may be a more powerful collision of the atoms, and more vibration, or heat.

(*f*) Chemical action and the production of heat, are connected more or less with electricity and magnetism. Our earth experiences magnetic influences from the sun; and the tails of comets, always pointing away from the sun, whether in coming or going, indicate an electric influence of great power at work in the sun. This may possibly add to its heat-producing power.

(*g*) There is an ether, perhaps many ethers, filling all space. By these a comet may be delayed in its orbit, and by these light, heat, magnetism, gravitation, and chemical influences, are constantly travelling to and fro, from star to star, in all directions. The precise connection of these with chemical action and the production of heat, is not yet known; but it is wonderful in how many ways they are associated with such action, and with the production of heat. The following is a most surprising instance: If between the poles of a very powerful magnet, a metallic cylinder be rapidly rotated, then that simple rotation produces an intense heat, sufficient to melt some metals. The heat seems to arise from the friction of the cylinder against an invisible ether collected there. The collection of such ethers around the sun is probable, at least possible; and these, along with the rotation of the sun, and the violent agitations on its surface and through its interior, may be connected with the production of heat.

(*h*) When a piece of lime is held in the flame of oxygen

and hydrogen, an astonishing flood of light is thrown off. A far greater amount of heat is also at the same time radiated. Other solids also, as platinum, wonderfully increase the light and heat. The sun is a solid, fused and incandescent in a flame; hence its vast amount of light and heat. The light we know comes chiefly from the body below the flame, and doubtless the heat has the same origin. We cannot account for all the light of a solid in the hydrogen flame on the theory of the conservation of force; neither can we, I think, account for all the radiant heat. It may be said that the heat gained by radiation is lost by convection, but this we do not know; to me it seems improbable, although the old theory points that way. But the sun can lose no heat by convection. And if a solid in a flame, whether fused or unfused, really increases the heat as it does the light, without detracting from the force in other ways, then we behold a condition in the sun favorable to the chemical theory. The theory of the conservation of force is yet in its infancy. It cannot embrace the force of gravity, notwithstanding the noble attempts of Faraday and others. Hence Mayer regards gravity as a property of matter, and not as one of the forces capable of conversion and reconversion. Newton, however, with great clearness maintained that it could not be regarded as a property of matter, but as a force acting mechanically on matter from without. But all the force of gravity constantly expended in the direction of the centre of our earth, is lost, as far as we can perceive or think. Force, therefore, may be annihilated, as far as we can learn or imagine; a proposition directly opposed to the theory of conservation of force. I I mention this, not to disparage the great acquisition to science of that theory, but to show how it is still in its crude beginnings, and not to be abused by undue applications as in the case of the sun, which is a solid fused within a flame. It is like the lime or platinum in the oxy-hydrogen flame, and how much light and heat it can send forth we do not know.

The force of gravity expended on the sun may be supposed to be converted into heat and light; but this is untrue, because no such heat and light come from the planets. All this discrepancy between gravity and the theory of conversion of force, shows what unexpected difficulties we may meet in the application of a new and valuable theory, and how such applications may lead us into error. The case is the same in applying the chemical theory to the sun. Scientific men could not account for the heat of the sun by chemical action, according to their favorite idea of a precise amount of fuel for a precise amount of heat, and therefore they give up chemical action altogether, not remembering that there are other ideas in the case as important as their favorite one—other ideas founded on possibilities, probabilities, and certainties which should have the controlling influence. It is not uncommon in morals, in religion, and in politics, as well as in science, for men to fasten their minds on a single idea, and to carry that idea out to the most mischievous consequences, until at length they, or the next generation, are awakened to find that there are other ideas to be taken into account far grander, more general, and more influential. The idea that by chemical action a pound of matter of any kind all over the universe, and under all possible physical conditions, must give only the same amount of heat as a pound of our charcoal burned in the open air, has been a very pleasing idea; but it is rashly theoretical, and especially mischievous when applied to distant parts of the universe millions of miles away, and to conditions so different and unknown as those in the sun. There are plenty of evidences of chemical action in the sun, but no evidences that we must carry our charcoal idea there.

Thus we perceive, from the eight foregoing items, that the conditions of chemical action in the sun are in some points certainly, and in others probably, very different from those in one of our little furnaces. These, conditions all relate to the heat-giving power; all tend to add force to chemical action and to the augmentation of heat. Therefore, the extraordinary heat-giving power of the sun should not lead us to deny chemical action in the sun, but to inquire in what ways and by what means chemical action in the sun produces so much more heat than on the earth. The true scientific course is not to deny chemical action in the sun, in the face of so many evidences in its favor, but to ascertain how chemical action is so greatly modified there, what extraordinary agencies are at work there, and how the sun, the great central body of our system, still alive with the play of all imaginable physical forces, differs from this small, cold, and dead globe, deriving every breath of air, all its movements, and all its appearance of vitality, from the sun. We must remember that we know very little of what chemical action really is. What makes two substances combine chemically, is still wholly unknown. How many different physical forces are at work in the act we cannot tell. Neither can we tell what is the combined action, what the grand and glorious play, of all the physical forces in all their might and majesty on the vast central and only active globe in our system. It would be most unscientific to suppose that chemical action there is in all respects the same as chemical action here. What are our small laboratories compared with the laboratory in the sun, nearly 3,000,000 miles in circumference? Most certainly chemical action there must be greatly modified, and in what those modifications consist, is our proper inquiry. In the next section I will endeavor to show that those modifications are so great as to produce those peculiarities in the condition of matter which we call simple chemical elements. These elements are all fundamentally the same, and we can greatly change them, put them into allotropic states, even by our inefficient means. In the sun, there are reasons for believing that the production of new elements has always been going on. Just as new species of plants and animals have from time to time been created on our earth, one assemblage succeeding another; so in the sun, and also in the earth during its former fiery condition, new species of elements have succeeded one another, and all created through the agency of physical forces from the same fundamental and essential matter. As all organized beings, plants, and animals, contain in their composition the same mother earth, using the same atoms over and over again, so the same essential matter has probably been moulded through many forms, many so-called simple elements; and the process is ended, we know not how, only when the activity of the physical forces ends in what we call a lost star.

SECTION XIII.

THE THIRD ASSUMPTION: THAT NO FUEL OR NO NEW CHEMICAL ELEMENTS CAN BE ADDED TO THE SUN.

In showing that this assumption is unwarrantable, I propose to give several facts in favor of the contrary idea, that new combustibles may now be prepared in the sun from materials which have already been burned. These facts I will draw partly from the nature of matter as revealed by Natural Philosophy and Chemistry, and partly from the history of matter as revealed by the Nebular Theory. The nebular theory I will assume to be perfectly true, such as I think I have proven in the remaining portion of this work. It regards all matter as having been nearly equally diffused through all known space, and nearly homogeneous or of the same nature everywhere. By a special process it has contracted and condensed into globes, and now I shall show that these globes have very different simple chemical elements; that these simple elements are mere modifications of one general fundamental matter; and that these modifications have been formed in the process of contraction and condensation, and therefore that they are still forming in the sun. The sun is eight times less dense than is required by the law of density in the solar system. Therefore, in its present active state of physical forces, it is still in the process of contraction, and will be so until its fires are burned out, and its due density shall be reached. We will now attend to these facts, or classes of facts, which I have proposed to give. They are eight in number:

FIRST FACT.—The condensed globes of space, the fixed stars, the sun, the planets, the satellites, and the meteorites, from having formerly been all of nearly the same kind of material, are now composed of very different simple elements.

The Fixed Stars. The dark fixed lines in the stellar spectra have led to the following conclusions:

The fixed stars differ in their simple elements more or less from the sun;

They all differ more or less from one another;

Some of them have a general resemblance to the sun;

Some of them have resemblances among one another;

Resemblances of fixed lines are accompanied, occasionally at least, with resemblances of color;

Changes of color seem to be attended by changes of fixed lines.

These observations, however, about the resemblances in the simple elements among the stars, and the connection between their colors and their fixed lines, must be regarded as only provisional, and in waiting for more thorough investigations. The great and undoubted result is that the spectra of no two stars are identical; and hence the infinite diversification of matter among the infinite number of stars. From being the same or nearly the same, it has received an inconceivable number of modifications.

The densities of the stars, as already mentioned, are of the same purport as their fixed lines. The very slow motions, and the want of sensible motion, in the companions of some double and multiple stars, indicate a lightness similar to our rarest gases; still they are not gases, for those stars have liquid bodies and envelopes of flame. Their elements are analogous to ours, but they are extremely rare, and hence extremely different from ours.

The Sun. Every successive improvement in the apparatus for observing the solar spectrum brings into view new dark lines. A recent observer says, it is as impossible to count the dark lines in the solar spectrum as to count the number of the stars. This strongly proves the great number of simple chemical elements in the sun, and leaves on the mind the impression of several thousands.

The Planets and Satellites. The densities of the planets and their satellites prove that they are composed of very different elements. Mercury is more than sixty times, and our earth about fifty times, more dense than the inner moon of Jupiter. Saturn is only about one-ninth as dense as the earth; it would float buoyantly on water. There is a high probability that the satellites of Saturn and Uranus are far lighter than those of Jupiter. Between the two extremes of the attendants of the sun, there is probably a greater difference in density than one hundred to one; and from one extreme to the other there are regular gradations of small amount.

The Moon. The difference in constitution between the earth and the moon is seen in their densities: that of the moon being about half that of the earth. The nitrogen of our globe is found only in the atmosphere, and such substances as derive it from the atmosphere. The moon has no appreciable atmosphere, and therefore, in a high probability, no nitrogen.

The Earth. Although there are about sixty-three chemical elements in the earth, yet one of these, oxygen, forms nearly or quite one-half of the globe; three of the elements, oxygen, silicon, and aluminum, form more than three-fourths of the globe; and if to these be added calcium and iron, they would form about seven-eighths of the globe. The other more abundant elements are carbon, hydrogen, nitrogen, chlorine, sodium, potassium, magnesium, sulphur, and phosphorus. While the metallic elements are far more numerous, being as fifty to thirteen, the non-metallic elements are far more abundant in quantity. Among the metallic elements there is greater sameness; among the non-metallic there is greater variety and a wider difference in qualities. The non-metallic form by weight more than three-fourths of the globe. This, of course, is an estimation of the entire globe by its solid crust. The idea that our globe was once in a gaseous condition, is confirmed by the fact that about half the globe, estimating by weight, is still gaseous at ordinary temperatures; that is, the elements to this amount are gaseous when not chemically combined.

While the difference in density between the extreme elements of our globe is as 256,700 to 1, there is still an easy gradation among them; and this easy gradation extends also to their other principal properties. We may so arrange them that in their review we shall find the differences very small. All the fifty metals, for instance, have many important properties in common. Some have very near resemblances, as potassium and sodium, calcium and magnesium, iron and magnesia. The practical chemist will readily think of many others. The transition between the metallic and the non-metallic elements is also very slight; indeed, it has been a matter of uncertainty to which of these divisions some few elements belong. Selenium and silicon have been classed among the metals, and carbon has some of the properties of metals. At least one of the metals, gold, has the property, when in thin leaves, of being translucent. Among the thirteen non-metallic elements there are also remarkable resemblances; as between sulphur and phosphorus, and between the family of the haloid elements, fluorine, chlorine, bromine, and iodine. In fact, the simple elements may be classified, the same as plants and animals are classified. They may be divided first into two great divisions, the metallic and the non-metallic; and these may be divided again into families, and the families subdivided into minor groups. This fact indicates near relationships in a genetic point of view.

The Meteorites. These, like the moon, are probably satellites of the earth; but being very small, they are liable to extraordinary perturbations, and hence strike the earth in many directions. Like the planets and satellites, their different densities indicate the differences of their simple elements. Some are as heavy as iron, others are as light as coal. Accordingly, in their chemical constitutions, they have been found by analysis to vary greatly among themselves. Some have ninety-six hundredths of iron in them, and others only two hundredths. In meteorites of all classes there have been found only eighteen simple elements; and these are the same as those in our earth. A frequent combination, that of iron and nickel, is peculiar to some meteorites, and they have a few other less conspicuous minerals peculiar to themselves. It is remarkable that the elements which are so exceedingly abundant in the earth, oxygen, silicon, and aluminum, are not the most abundant in meteorites. Oxygen is sparingly present, and not enough to oxydize the iron and nickel. They have no hydrogen, nitrogen, or chlorine. The greater satellite, the moon, exhibits a similar want of gases. It is airless and oceanless, and seems to be a great rocky mass, like a huge meteorite.

Thus while matter was once everywhere the same, and while it is still everywhere *fundamentally* the same, it has nevertheless been modified in an infinite number of ways, which modifications we call simple chemical elements. The former equal diffusion of matter through all known space, and the uniformity of its properties wherever diffused, lie at the foundation of the nebular theory. And it has been received as an elementary truth that matter, in its essential and fundamental properties, is still through all known creation identical. In all the globes of space we see matter alike subject to gravitation, inertia, and the laws of motion; in them all it possesses the property of existing in the three states, solid, liquid, and gaseous, and therefore it must be alike subject to the atomic force of attraction and repulsion. In them all it is subject to be modified into different simple chemical elements. But superadded to this oneness of matter, we behold in every fixed star, in the sun, in the moon, in the earth, in all the planets, in all the satellites, in all the meteorites, an endless number of the modifications of matter. There are no two great globes alike in simple elements. There have been causes in operation which have produced an infinite diversity. In all this we recognize a familiar principle in creation. It is diversity in unity. There is a unity in all vegetables, but how wonderfully is the idea of a vegetable varied in more than a hundred thousand different species! Even in the same species we see no two precisely alike. Therefore we conclude that since the former equal diffusion of matter, and during the long process of its aggregation into globes, it has been subject to physical forces which have produced those modifications called simple chemical elements. As every oak grows somewhat differently from his fellow, and as every difference has had a real cause which we seldom understand, so every globe of space has grown in its chemical elements to be somewhat different from its companions, from causes in the operation of physical forces which we need not now stop to examine. All rivers, islands, and mountains, have been formed by the operation of the same forces, but no two rivers, no two islands, no two mountains are alike. In the same manner, the infinite diversity among the chemical elements, formed by the operation of general physical forces, should give us no surprise. And as among the various species of plants and animals, we see a gentle gradation from one to another with here and there a gap of moderate size, so among the simple elements in our own globe we find similar gradations, and these gradations broken occasionally by unexpected, though moderately sized gaps. Still in a general way the maxim holds true, "Natura non facit sultum." It has already been remarked how the elements may be divided and subdivided after the manner of the ordinary classifications among plants and animals. By two methods entirely independent it has now been shown that all matter has the same general primary properties, and that it has received various modifications which may be called secondary properties. The one method is by the nebular theory, which assumes that all matter was once universally and equally diffused, and consequently everywhere the same. In the course of its aggregation into globes, it has received very different secondary properties. The second method is by direct observation on the heavenly bodies. By this we perceive that they have now certain primary properties which are the same in all, and that each one, as nearly as our observations can go, has its peculiar secondary properties. There is a third method, far more convincing, and which gives the other two nearly all their force. It is the method of examining the matter of our own globe with all the instruments and with all the light afforded by modern science.

SECOND FACT.—This brings us to our Second Fact, which is to show that the primary properties are much more numerous and important than the secondary; that the secondary properties are really very slight, and such as may be the result of known physical forces. Nevertheless the secondary properties give to the material world all the strange variety that meets our every sense.

The general properties of matter are these—fourteen in number, according to my examination:

1. Divisibility into atoms which to us seem infinitely minute and ultimate.
2. Indestructibility; however much it may be divided and subjected to -powerful forces, none can be annihilated.
3. Occupancy of space, a property formerly called impenetrability.
4. Capability of existing in three very different conditions; gases, liquids, and solids. Some few of the gases, when alone, have not yet been liquefied or solidified, but this can be done by bringing them first into combination.
5. Inertia, the property of offering resistance to force, whether in rest or in motion.
6. Capability of motion, according to definite laws, on the application of the forces to which it is in subjection.
7. Subjection to mechanical force.
8. Subjection to the force of gravitation.
9. Subjection to the force of magnetism.

10. Subjection to the force of electricity.
11. Subjection to the force of heat.
12. Capability of producing certain actions on light and heat, as reflecting, refracting, polarizing, and absorbing.
13. Subjection to atomic force, as in the attraction of cohesion between the particles of water.
14. Subjection to the chemical force, as in chemical attraction and repulsion.

All matter possesses these fourteen general and fundamental properties. We may therefore regard it everywhere and under all its modifications as being essentially the same. The difference between the various modifications of matter, the simple elements, *consists not in the possession of any new properties, different from the above fourteen, but entirely in the degree and manner in which they possess the last six of the fourteen general properties.* This I regard as a highly important truth, and deserving illustration. All modifications of matter—simple elements—are affected by the forces of magnetism and electricity, but some more and others less; some in one way and some in another. All are in subjection to the atomic force, but in different degrees; some are held firmly by this force, and are made strong, as iron; others are held by it feebly, and are made soft and tender, as sodium; some it makes very dense, as gold; others very rare, as aluminum. This same force, by different degrees and modes of operation, imparts to various bodies quite different conditions, as those of elasticity, brittleness, porosity, ductility, malleability, and the like. While all matter possesses the primary property of being in subjection to the chemical force, some elements are affected by it in one way, and some in another. Again, all the simple elements are affected by light and heat, and reciprocally they act on light and heat, but in different degrees and ways.

Since all matter possesses these fourteen fundamental and primary properties, and since some portions of matter differ from others merely in the degree and manner in which they possess the last six of these fourteen properties, it is evident that these differences, these secondary properties, are comparatively slight. These comparatively slight differences in different portions of matter are called the simple elements; they are evidently mere circumstances of manner and degree, and I think I can prove conclusively that they have been produced by the operation of physical forces during the condensation of the nebulous masses. But first, let us see what we can do by artificial means.

THIRD FACT.—By our feeble handling of the physical forces we can produce some modifications in matter very nearly, if not precisely, the same as the natural simple elements. The chief difference is that our artificial modifications are not so permanent as those which are natural, and this may be accounted for by the smallness and feebleness of our instrumentalities. We can also alter and modify the peculiar properties of some of the natural simple elements. A few instances may here be given:

Oxygen perfectly dry and pure may be acted on by electricity while passing through a glass tube, and when it comes out it is ozone. Ozone has new properties, which oxygen has not. It differs far more from oxygen than some of the simple elements differ from others; many metals, for example.

Chlorine, when acted upon by the rays of the sun, chiefly the indigo rays according to Draper, of New York, receives a new property quite different from chlorine prepared and kept in the dark. When prepared in the light, it acquires the power of decomposing water, even when put in the dark afterwards.

Iron may be put in a condition in which, like gold, it

resists the action of nitric acid, although in its ordinary state it is very readily dissolved by that acid.

Carbon may be changed into several conditions which have very different properties. The diamond is transparent and hard. Lampblack is directly the opposite. In all other forms carbon is highly combustible, but when in the condition of plumbago it is used as crucibles to resist the action of fire. This property cannot be attributed to a very minute mixture of iron, for iron is also combustible. The internal and molecular differences between the several forms of carbon are proved by their differences of specific heat.

Ammonium. As there are evidences of the existence of the simple element fluorine, although that gas has never been separated, so there are evidences of the existence of the metal ammonium although it has never been separated from its compounds. This metal is artificially produced. It is composed of nitrogen and hydrogen. In all its relations to the other simple elements and to the laws of chemical combination, it comports itself as a true simple element. Hence we have chloride of ammonium, and sulphydret of ammonium, and the like.

Cyanogen is composed of nitrogen and carbon, and in all chemical compounds and chemical processes it sustains the relations and peculiarities of a simple element. Hence it is called the compound radicle. The fixed lines in its spectrum, I have been informed, also give it the character of a simple element.

Schonbein, after many experiments, and an investigation during six months, arrived at the conclusion, that the simple elements chlorine, fluorine, iodine, and bromine, are really compound substances. He gave an extremely interesting account of his proceedings in a letter to Faraday, which was published in the "Philosophical Magazine," May, 1859.

The four examples of oxygen, chlorine, iron, and carbon, are enough to show how those modifications of matter which constitute simple elements, can be still further modified by the ordinary physical forces. New properties, such as characterize simple elements, can be imparted to them. The examples of ammonium and cyanogen, show that substances which are really compounds may act like simple elements in all their chemical relations. Such facts tend to show that these characteristics which give to bodies the appellation of simple elements, are but modifications brought about in former epochs, by mighty forces which are now in operation in our hands in a very small way. On account of this smallness, the modifications which we produce have not the permanence of those produced in the former nebular masses. There is a great difference between our electric sparks and a clap of thunder. And there is a far greater difference between our little apparatus and the great nebular laboratories, many millions of miles in diameter. To these we will now attend.

FOURTH FACT.—Before our earth had assumed its present condition, there are evidences that it was so highly heated as to have been entirely melted. Other evidences lead to the belief that in a still anterior period, it was in a gaseous or vaporous condition, and that in a yet previous period, it formed, along with the sun, and all the other planets and satellites, a vast nebulous fiery mass. Leaving the demonstration of this for another occasion, we must here assume the Nebular Theory as true, and, according to this, all known matter was once gaseous or nebulous, universally and equally diffused through space, and by condensation it received the forms of fixed stars, suns, and their attending bodies. The grand inquiry now occurs, What caused this condensation? I can think of no other cause than chemical action. Whenever chemical combination takes place between two or more bodies, the resulting compound always occupies less space. This is the invariable law. The explosion of gunpowder is no exception, for this is at first a decomposition and a consequent vast expansion; afterwards its elements combine again, but in a different way, so as to occupy much more space than when in powder.

It has hitherto been supposed that the contraction of the former nebulous masses was due to their loss of heat. That great expansion of matter, it has been assumed, was caused by intense heat, and the constant radiation of this heat away in the realms of space, left these masses constantly cooler and smaller. But there are two impossibilities in the way of this supposition. First, we can think of no cause for this intense heat. Heat can no more exist without a cause than motion; and for heat on so grand a scale through all space, and in such intensity, we cannot possibly conceive of any cause. Secondly, if all space were thus heated, there could be no possibility of its radiating away. It must forever, as far as we can see, remain the same.

But if matter were originally homogeneous, or so nearly homogeneous as to have but two or three modifications, then if chemical combination began, there would be a simultaneous contraction; there would be a breaking up of the original mass into separate nebular masses; there would be intervening vacant spaces; there would be the production of heat and light; and then would begin the radiation of that heat into the spaces left vacant. As this chemical combination should go on, the open spaces would become larger, and the material masses would grow smaller, until they arrived at their present sizes. The fiery condition would all the while be maintained until, as in the case of the earth, and more especially the moon, chemical equilibrium would be attained, and then the fire and the luminosity would die out.

But what caused this original chemical combination? This question we will be able to answer as soon as we learn what causes chemical action now. We do not know what causes oxygen and hydrogen or oxygen and carbon to combine and give out heat and light. This we may one day discover, and then we will be prepared for the question respecting the beginning of the first chemical combination. We are at liberty to assume that matter in this state of universal diffusion had only two or three modifications, and that all others have been derived from these by different combinations; because we know how many dozens of different substances may be formed from these three, oxygen, hydrogen, and carbon. The same atoms, combined in different ways and in different proportions, form different compounds.

As a further elucidation, I offer the following: The universal diffusion of matter must have been caused by an excess of the force of repulsion between the several atoms almost infinitely beyond what exists now among the gases. It must at least have counterbalanced, and a little more than counterbalanced, the force of gravitation. This great force of repulsion must have been converted by some means, as yet unknown, into the force of chemical action. This conversion of the forces is now a familiar idea. Heat may be converted into motion, and motion into heat. The electric force may be converted into chemical force, and the chemical force into the electric force; and similarly between the other physical forces. According to this idea, the force of repulsion, which once held all matter equally diffused, completely counteracting universal gravitation, is now expiring in the form of chemical action, in one after another of the lost stars—not really expiring, but changing into heat and light.

In those great nebulous masses which formed our solar

system, and the many systems of the fixed stars, we see as many chemical laboratories, in operation thousands of millions of years, on a grand and magnificent scale. Our own system once occupied a space many times larger than the orbit of Neptune. All its materials then were many million times more rare than hydrogen, which is even yet a quarter of a million times more rare than platinum. The play of all the physical forces during so long a time, in such immense bodies, and in such vast contractions from their extreme rarity to their present densities, is entirely beyond our conceptions. But here we behold a state of things and a scale of operations exactly suited for the modification of matter. If we with our small manipulations can produce striking changes, such as those we have detailed in oxygen, chlorine, iron, carbon, ammonium, and cyanogen, what may we not expect in a condensation from such a rare condition during so long a time, and under the action of all the physical forces known and unknown? The unknown physical forces are probably numerous. It is but a little while since the discovery of the force of gravitation, with its remarkable laws and universal extension. Electricity, magnetism, and galvanism are quite recent discoveries. A chemical medium of universal extension, and vibrating in the starbeams, is only just now breaking on our view. The vibratory nature of light and heat, and the forces residing in each, are among the latest wonders of science. All these physical forces must have been in active operation during the condensation of the nebulous globes, and how many more physical forces were engaged in the great work we cannot imagine. Neither can we tell what were those great chemical operations; evidently our little laboratories can give us no just ideas. But among those forces known and unknown, and in those immense and long-continued operations, in these condensations so extremely great, it is reasonable to suppose that modifications of matter took place which are permanent, not temporary, like those which we can make. The modifications which we can make, we can alter again and destroy; but we need not wonder that we cannot, with our small means. alter and destroy all the modifications made by such celestial means, during such periods, and in such condensations. Seeing, therefore, what vast, what mighty, and what long-continued laboratories have been at work everywhere through space, we should not be surprised that matter in all the great globes of space has been modified in an infinite number of ways. The theory of chemical action, therefore, accounts not only for the light and heat of the celestial bodies and for their changes, but it accounts also for the origin of the simple chemical elements, and for the fact that in all the great globes of space the elements are different.

There are two ways in which we can conceive that new elements may be formed. The one is by a change in some previous element, as the change from oxygen to ozone. It has been happily said that the difference between ozone and oxygen consists in a different packing of the atoms. Because our packing is not permanent, that is no reason why the packing during the nebulous condition, under mightier forces, should not be permanent. The other conceivable way of forming a new element is by compounding previous simple elements. This would, indeed, make a really compound element, but it would be simple to us as long as we are unable to decompose it. Probably every intelligent chemist suspects that what we call simple elements are not truly simple, but only undecomposable as yet by our means. The mighty forces at work in the sun—and they are inconceivably mighty, judging from the light and heat—can undoubtedly put things together which our little means can never take apart, especially during those long periods for work while the sun was condensing from beyond the orbit of Neptune to that of the earth. To all this must be added the powers at work in the earth during the long and vast condensation from a distance far beyond the orbit of the moon. Therefore, what we call simple elements may be not only compounds, but double compounds, or compounds that are the result of hundreds or thousands of compoundings. In this way there must have been at different periods entirely different sets of elements in the sun; the same as at different periods there have been different sets of plants and animals on our globe. For this reason the chemical action in the sun may have continued so long; after one set had combined, another set has been ready for a like operation. Further probability is lent to this frequent compounding in order to reach the present results, by the infinite minuteness of the ultimate atoms of matter, and by the wide diffusion of the materials of the sun in a nebulous condition, being some hundred million times more rare than hydrogen, which is, as just said, still a quarter of a million times more rare than platinum. Each compound diminishes the space occupied by its previous simple elements, and judging from the vast diminutions of space occupied by matter, we are led to think of an almost infinite number of compoundings, each one following its predecessor. Remembering the many wonderful illustrations of the infinite minuteness of the atoms of matter and also their former wide diffusion through space, we are astonished to think how much farther apart these atoms then must have been than they are now. Even now the atoms are farther apart than their diameters. The intervening spaces, then, must have been filled with these media, which are the instruments of force, then existing as force in reserve. Each new compounding and packing must have expended some of that force, and must have brought the atoms closer and closer together, and therefore the series of changes in matter, during millions of centuries, must have been marvellously numerous. The inquiry, What becomes of the products of combustion in the sun ? is in this view easily settled. Those compound products may combine with one another, the same as some compounds here may combine and give out heat and light, only those in the sun are on a larger and more intense scale. There are still several other astronomical facts which prove that the simple elements were created during the fiery nebulous condensations. These will here be added :

FIFTH FACT.—The densities of the several planets, and consequently the nature of their simple elements, are such that these elements must have been formed since their separation from the sun. Taking water as a standard of unity, their densities are as follows : Neptune, 1.25 ; Uranus, .97 ; Saturn, .76 ; Jupiter, 1.32 ; Mars, 5.21 ; Earth, 5.44 ; Venus, 5.11 ; Mercury, 6.71. These figures are from Humboldt's "Cosmos," and consequently they are German authorities. Herschel's "Outlines" makes Neptune equal to Saturn. With this amendment the densities of the several planets increase generally in the direction toward the sun. This is accounted for by the Nebular Theory. The outer portion of a nebular mass must always be the lightest. The planets were separated in the order of time from the outer to the inner. Hence their densities should strictly follow that law ; every one should be lighter than its next inner neighbor. But there are two striking exceptions, Saturn and Venus. Instead of being heavier, they are lighter than their next exterior neighbors. Once they were more dense, now they are less dense. This change has occurred during their separate nebular periods. It is a change in the elements of the planets, because density depends on the nature of the elements. The explanation is easy and natural. In the Earth, the planet next exterior to Venus, a set of simple elements has been formed, denser than those in Venus, out of material which before was lighter. This was done during the process of nebular condensation. The same must be said of Uranus, the next exterior neighbor to Saturn; and also of two of Jupiter's satellites. The first or nearest is less dense than the second, and the second is less dense than the third.

SIXTH FACT.—Another fact, proving that the present elements of the planets have been created since their separation from the sun and during their fiery nebulous histories, is seen in the nature of the elements of our earth. Their differences in densities are so wide that they could not have existed in the same layer at the surface of the sun, when in a nebulous condition. Not only platinum but also gold and several other metals are nearly a quarter of a million times more dense than hydrogen. They are also enormously more dense than the other gases, or than the vapor of water. The amount of these metals is considerable, for they are very generally diffused through the crust of the earth. In the clay beneath this city, from which its bricks are made, we have recently been told that the quantity of gold is large. There is enough in the walls of every house to gild their entire outer surfaces. There must indeed have been vertical currents in the sun even in its nebular condition, and consequently a mixing up of the heavier with the lighter materials to a certain extent. But there could have been no mixing gold with hydrogen, or even iron with the vapor of water. Had these elements then existed, the gases and the vapor of water would have been far away on the surface of the atmosphere of the sun, and the gold and iron down in the interior. In the extremely thin nebular ring which separated from the sun and formed our earth, the elements must have been not far from the same density, and their wide differences must have originated during the vast chemical operations which allowed the earth to condense from far beyond the orbit of the moon to its present size.

SEVENTH FACT.—The composition of Meteorites is another fact tending to prove that the simple chemical elements have been formed during the process of nebular condensation. They have only eighteen simple elements in them all, and none of these are different from those in our earth. According to the nebular theory, they must have been given off from our earth by centrifugal force while in a nebulous condition, either before or since the parting of the ring which formed the moon. Being very small, we cannot suppose that their interior chemical action proceeded with all that power which belongs to great bodies like the sun or the earth, and which is necessary to form new elements. Hence no new element was formed within them after parting from the earth. But in the great laboratory of the earth, the forces were at work on a much larger scale, and during a much longer continuance, and hence they wrought out more than forty additional elements after the parting of the meteorites. Thus the chemical theory accounts in a very natural manner for these two circumstances: first, why there are so few simple elements in the meteorites; and, secondly, why none of them are different from those in the earth. It may be said that the meteorites should be regarded as small asteroids circulating around the sun exterior to the earth's orbit, and not as satellites revolving around the earth. This supposition would lead to the same explanation of the elements of meteorites. The deficiency of the gaseous elements in meteorites may be explained by the supposition that these had not yet been formed, or had been only sparingly formed. Oxygen, for instance, was so deficient, that the iron and the nickel are not oxydized.

9*

EIGHTH FACT.—In the sun there are two facts which favor these views of the formation of the simple elements during the process of condensation—its want of density, and its active display of the physical forces. The density of the sun is only about one-fourth of the density of the earth. On two separate grounds we should have expected to find the sun at least ten times more dense than it now is: first, because of the great amount of its own gravity, compressing its volume with a power far beyond that of the planets; and secondly, because of the law of density of the solar system. This law, in its pure generality and completeness, has never yet been stated, that I am aware; it is this: The central parts of the solar system are the more dense, the circumferential parts are the more rare. The illustrations are of three quite independent classes: First. The several layers of the earth and of the planets are more rare at their surfaces and more dense as they approach their centres. Secondly. The planets being central in their own systems, are more dense than their satellites. Thirdly. The planetary members of the solar system generally increase in density from Neptune, at the circumference, to Mercury, the most central. The sun is the great exception to this law of his own system, the very member where the law should most dominate; and on account of this fact, and of his exceeding force of gravity compressing his volume, we must look upon his want of density as being due to some great and noteworthy cause.

This cause we cannot suppose to be its expansion by heat; for heat never expands a liquid or solid ten times its volume—it flies off into vapor before that. The only supposable cause is, that the sun has not yet received such a set of simple elements as to make it ten times more dense than it now is. But there the physical forces are still in an inconceivable state of activity. By their action a set of elements may be prepared which is to impart the due degree of density. Therefore, the two facts, that the sun is not yet dense enough to follow the law of the solar system, and that in its constitution the physical forces are still active, agree well together, and bring the sun in harmony with the law of his numerous attendants. In the planets and satellites, the activity of the physical forces has ceased, and their densities are according to law. In the sun the activity of those forces has not yet ceased, and therefore its density is not yet according to law. It has not yet received the elements required for its due density. They are to be elaborated during the sun's condensation. When that process is completed according to the law, his diameter will be less than half of what it is now.

The eight facts, or classes of facts, now reviewed in this section, seem to me to be exceedingly strong to prove that the simple elements have been formed during the period of nebular condensation, and by the agency of physical forces employed in the process of chemical combination—a process not yet in the least understood.

Matter nearly equally diffused through space, must have been nearly homogeneous; but after its condensation into globes, we find that all these globes have different simple elements. No two fixed stars have the same lines in their spectra. The earth was once less dense than Venus, but during its condensation it received a set of simple elements which made it more dense. The same change of relation took place between Uranus and Saturn. When the earth separated from the sun in the form of a nebular ring, its materials must have all been so nearly equally dense as to float together on the surface of the sun's equator; but now after a long nebular condensation, a new set of elements have been formed, some of which are about a quarter of a million of times more dense than others. The earth had once only about eighteen simple elements, and they were not different from those now in the meteorites; but during nebular condensation about forty new elements have been formed. The sun, to agree with the law of density of the solar system, should be ten times more dense than it is now; but this want of density can be accounted for by the fact that the physical forces are still at work in its constitution with the most inconceivable activity, and therefore they have time enough to elaborate a set of elements of due density. These simple elements are in all cases mere modifications of an original and fundamental matter. This is made evident by the facts that they all have fourteen essential properties in common, and that they differ only in the manner and degree of the last six of these properties. Such modifications of matter as simple chemical elements, we can make by our feeble employment of the physical forces; though on account of this feebleness our modifications are not permanent. But in the condensations of the great globes of space, condensations reducing their volumes millions upon millions of times, alive all the while with the active play of the physical forces, it is reasonable to suppose that these forces in such vast bodies and during such long periods, could produce modifications in matter which to our imperfect means must appear as simple undecomposable elements.

Evidently at the present period the activity of the physical forces in the sun, is beyond all our powers of conception; and as the chemical elements have originated during the past activities of the stars, of the sun, and of the earth, it is unscientific to suppose that the production of such elements has ceased in the sun, unless we have some reason for such a supposition. What has been going on we must believe is still going on, until some evidence appears for a cessation. In the earth an equilibrium is attained among its forces, and all is still, save the impulses which come from abroad. No more modifications of matter here occur, except such small ones as our feeble means can make. But in the sun the case is very different. There the formation of new chemical elements may be going on as rapidly as ever. Simply by admitting this, we then must yield our belief to the vast number of facts in the earth, in the sun, and in the stars, which go to prove that their great changes in light and heat, as well as their continuance and discontinuance in light and heat, are the natural results of chemical action.

I have now finished what I have to say on the only objection ever raised against the theory that the light and heat of the sun are produced by chemical action. I have shown that this objection—the want of fuel in the sun—is unfounded; for it proceeds on the three unwarrantable assumptions. In opposition to the first assumption I have shown that it is possible, and even probable, that the materials of the sun, pound for pound, give out far more heat than those in our earth. In opposition to the second assumption I have shown that it is possible, and even probable, that in the vastly different conditions of chemical action in the great central orb of our system, more heat may be given out by the very same materials than in our small furnaces. In opposition to the third assumption I have shown that it is possible, and even probable, that by the action of the physical forces at work in the process of chemical action and condensation, new fuel may be constantly prepared to keep alive the burning in the sun.

Therefore regarding the objection as invalid, and not warranted when we take a wide view of facts, we are compelled to fall back on the many positive arguments that our own globe was once in a blaze by chemical action; that the sun is now in such a blaze, and also the fixed stars; and so they will remain until, like our earth, their lights go out, and they are called lost stars. In view of all the evidences in favor of chemical action in these great stars, and also formerly in our earth, the true scientific course is not to deny that action on account of any apparent difficulties, but to inquire how it differs from that action in our small laboratories; how it is modified by the action of physical forces known or unknown. The play of these forces in condensations many thousand millions of times, must have been powerful beyond conception. We are terrified at the action of electricity in the condensation of a mere summer cloud; the rolling thunder, the vivid lightning, the intense heat, and the mechanical power, affright us; but what is the shower-cloud compared with the condensation of a world? We know that the action of the physical forces is such as to form new elements, for it is beyond doubt that these elements were formed during the process of nebular condensation. These elements, in their turn, are new sources of light and heat, and so the operation goes on until finally the great clock of forces runs down in one after another of these globes, and they become unluminous and cold. About the origin of these forces, and the primal winding up of these great clocks of eternity, we do not now inquire.

SECTION XIV.

THE FACTS EXPLAINED BY THE CHEMICAL THEORY OF STELLAR LIGHT AND HEAT.

The additions which the chemical theory of stellar light and heat, make to the acquisitions of science, and the confidence it should inspire in its truth, may be seen in the following large catalogue of facts for which it accounts:

I. THE FIXED STARS.

1. The chemical theory accounts for the exact and punctual periodicity of some of the periodic stars.

2. It accounts for the want of exact punctuality in the periods of others.

3. It accounts for the irregular routines in the waxing and waning of their light.

4. It accounts for the substitution of one irregular routine for another.

5. It accounts for the shorter time in the increase than in the decrease of the light of periodic stars.

6. It accounts for two maxima and two minima in one period.

7. It accounts for the variations in the maxima and minima, they being sometimes larger and sometimes smaller.

8. It accounts for the fact that the more constant the routines of waxing and waning in any star, the more punctual is its period; and the more inconstant its routines, the more unpunctual its period.

9. It accounts for the irregular stars.

10. It accounts for the lost stars.

11. It accounts for the temporary stars.

12. It accounts for the great fact that the invariable, the periodic, the irregular, the lost, and the temporary stars, graduate from one to another by insensible degrees.

13. It accounts for the endless variety of color among the different stars.

14. It accounts for their changes of color.

15. It accounts for the fact that the fixed lines are dark instead of variously colored in the light of the stars.

II. THE SUN.

16. It accounts for the simple chemical elements in a state of vapor held in suspension in the flames of the sun.

17. It accounts for the origin of the clouds in the sun, and their wonderful extent both horizontally and vertically.

18. It accounts for the invisibility of these clouds except in total eclipses.

19. It accounts for the existence of the flames.

20. It accounts for the agitations of the flames and their tall flakes or *faculæ*.

21. It accounts for the great and sudden changes in the shapes of the spots.

22. It accounts for the liquid condition of the body of the sun.

23. It accounts for the agitations and local currents in that liquid.

24. It accounts for the existence of the pores.

25. It accounts for the origin of the spots from the pores.

26. It accounts for the dark central nucleus in the spot.

27. It accounts for several nuclei occasionally in the same spot.

28. It accounts for the occasional existence of several shades of color in the same penumbra.

29. It accounts for the increasing and decreasing in the size of the nucleus as the penumbra increases or diminishes.

30. It accounts for the general fact that the penumbra appears before the nucleus, and shows, why it remains after the nucleus has vanished.

31. It accounts for the occasional exceptions to this general fact.

32. It accounts for the light border around the spots.

33. It accounts for the occasional bright streaks of light in the nuclei.

34. It accounts for the occasional broad patch of bright light in the central part of a spot.

35. It accounts for the breaking of the spots into fragments.

36. It accounts for the rapid flying asunder of the fragments after the breaking of a spot.

37. It accounts for the occasional approaching together or the separating of two spots.

38. It accounts for the fact that the rotation of the sun on its axis appears more rapid when estimated by some spots than when estimated by others in the same latitude.

39. It accounts for the more obscure and undefined appearance of the spots as they are vanishing.

40. It accounts for the final extinction or melting of the spots.

41. It accounts for the fact that the spots are not generally seen near the border of the sun.

42. It accounts for the nearer approach to the border by the larger than by the smaller spots.

43. It accounts for the apparent notch once said to have been made in the border of the sun by a spot.

44. It accounts for the fact that the eastern border of the penumbra is seen first, then the nucleus, and, lastly, the western border; and that when the spot is disappearing by the rotation of the sun, the eastern border of the penumbra first disappears, then the nucleus, and, lastly, the western border.

45. It accounts for the fact that the light of the sun is not polarized.

46. It accounts for the great heat of the sun.

47. It accounts for the light of the sun.

48. It accounts for the long continuance of the light and heat of the sun during so many millions of years, as revealed by geology.

49. It accounts for the want of density in the sun.

50. It accounts for the brighter appearance of a spot than of a planet in its transit across the sun.

51. It accounts for the radiation of heat by the spots, and that this heat is less than that from the flames.

III. THE EARTH AND THE MOON.

52. It accounts for the present general combination of the chemical elements of our globe.

53. It accounts for the partial combination of some of these elements.

54. It accounts for the want of combination between some of these elements.

55. It accounts for the present interior heat of the globe.

56. It accounts for the former entire fusion of the globe.

57. It accounts for the cessation of the former causes of heat, and for the consequent cooling of the globe.

58. It accounts for the great geological features of our globe in so far as they are the consequence of its cooling, such as the elevation of continents, islands, and mountains, the depression of the ocean beds, and the occurrence of volcanoes and earthquakes.

59. It accounts, by the same reason, for the volcanic and igneous structure of the moon.

60. It accounts for the extinction of the igneous action in the moon.

IV. THE COMPARISON OF FIXED STARS, SUNS, PLANETS, AND SATELLITES.

61. It accounts for the parallelism in the history of the great globes of space, the fixed stars, the sun, the planets, and the satellites. The parallelism appears in this way: The fixed stars are undoubtedly liquid in their constitutions. We have evidence of their cooling and solidifying in the irregular spots causing their periodicity, and in the total extinction of their light. We see the beginnings of cooling and solidifying in the pores and spots in the sun. The earth bears evidence of having been liquid, and of having cooled very gradually. The igneous structure of the moon, and the extinction of its volcanoes, indicate a similar his-

tory. It is the nature of the chemical elements, of which all these bodies are composed, to unite and produce heat: it is equally the nature of these same elements, after their union, to cease producing heat, and to allow these great bodies to cool, to lose their luminosity, and to solidify first on their exteriors in a thin crust, and afterwards very slowly in the interiors. All this remarkable parallelism in the huge globes which roll afar and near, is accounted for by this theory.

62. It accounts for the fact that in this history of the great globes, the satellites take precedence of the planets, and the planets take precedence of the sun, in losing their luminosity and heat. The moon is evidently in advance of the earth in cooling, because her volcanoes are extinct; and the earth is in advance of the sun, because it has lost its luminosity. After the separation of the moon, the earth had to contract, by chemical action, all the distance from the orbit of the moon. After the separation of the earth from the sun, the latter body had to contract by chemical action nearly two hundred million miles. In each case the greater amount of contraction must cause delay in ultimate cooling.

63. It accounts for two methods in the extinction of the light of a burning globe: first, by the gradual dying out of its fires equally all around; and secondly, by the formation of dark spots, which slowly increase, and at length envelop the entire star. The extinction of the temporary stars are examples of the first method, and the periodic stars are examples of the second. Upsilon Geminorum is visible only about one-eighth of its period, and Mira Ceti about one-third.

V. THE CHEMICAL ELEMENTS.

64. It accounts for the origin of the simple chemical elements.

65. It accounts for the fact that no two great globes of space are the same in their simple elements.

66. It accounts for the fact that interior planets and satellites are in a few cases less dense than their next exterior neighbors.

67. It accounts for the existence of such diverse elements in our globe, as hydrogen and platinum. Their existence together in our earth, without this explanation, would be fatal to the nebular theory.

68. It accounts for the contraction of the former nebulous masses.

69. It accounts for the fact that by the use of proper forces we can make such remarkable modifications in the simple chemical elements, as their allotropic states,—such modifications being merely examples on a small scale of those greater modifications which constitute the chemical elements.

70. It accounts for the fact that there are only eighteen simple elements in all the meteorites.

71. It accounts for the fact that the elements of the meteorites are not different from those in our earth.

There is a wide difference in their natures between many of these facts. A theory closely linking together such far distant diversities, and shedding the information we demand upon all, stands before us arrayed in strong recommendations. The problem before us is not simply to explain the cause of the light and heat of the sun and the fixed stars, but also to explain the partial and then the total extinction of their light and heat. Whatever theory we adopt, it must tell why the sun and many of the stars are bright and radiant with heat, and why the earth and many other stars have lost their luminosity. It must explain two different modes for the extinction of a star's light: one by the gradual dying out of its light, and the other by a periodical disappearance and reappearance ; its light at each reappearance growing less until it reappears no more. The theory must explain a vast number of important activities in the fixed stars ; such as their changes of color, and their several curious ways of varying the amounts of their light, and the appearance and disappearance of a temporary star. It must explain the activities in the sun, such as the spots, the pores, the clouds, and the faculæ or ridges of flame. It must explain why a fixed lost star has been left in the condition in which we now find the earth and the moon. It is wonderful what a wide range of phenomena must all be brought together and explained by one and the same theory. The chemical theory does all this, according to very familiar laws, easily and simply.

This theory is indeed remarkable for its simplicity, for its harmony with the known laws of the material world, and for revealing more clearly and more extensively than ever before the principle of unity between all parts of creation near and remote. It assimilates more than ever the earth and the great celestial bodies. This identifying more and more the earth and the stars was at the foundation of the two great advancements in modern astronomy by Copernicus and by Newton. Copernicus boldly made our earth, not the centre of the universe, around which the whole heavens revolved, but a planet like the other planets, wheeling around on its axis, and wheeling around the sun. Newton still further assimilated the earth and the stars. He extended an earthly principle to the sun and to the planets, and he showed how they are all alike actuated by the familiar force of gravitation. Later discoveries have extended the same principle to the fixed stars. The chemical theory binds the earth still more intimately with the sun, the planets, and the fixed stars. It shows that their natures and their histories are the same." And it dignifies the chemical force equally with the force of gravitation, by extending both alike throughout the entire known universe.

SECTION XV.

THE THEORY OF THE FORCE WHICH PROLONGS THE HEAT AND LIGHT OF THE SUN AND THE OTHER FIXED STARS.

It may be thought that I should have begun this Second Part by a formal statement of my theory of the force which has given so prolonged a duration to the light and heat of the sun, as is revealed by geology. But all theories should be founded on facts, and I preferred to bring forward the facts first. These facts, on account of the present condition of scientific opinion, I have thought best to present as rebutments to the one familiar objection against the chemical theory. Thus it comes out ultimately that the chemical force, now active in the sun, is the conversion or conservation of the atomic force of repulsion which once held the solar system in a nebulous condition. That condition was one of inconceivable rarity—many thousand billion times more rare than hydrogen, as will be shown in my demonstration of the nebular theory. The force requisite to expand the sun so far, must have been inconceivably great. That force, being indestructible, must still exist. But where has it gone? The answer is obvious: it is now passing off, and has long been passing off, as light and heat, through conversion into chemical forces. How the forms of force are converted into one another we cannot tell. We know not how mechanical force can be converted into heat, still we know that such changes can occur. But why suppose that the original repulsive force is converted into light and heat through chemical force, rather than through the electrical or some other mode of force? The answer is, because the present action in the sun, and in the fixed stars, and the former action in our earth, as already detailed in the beginning of this volume, all strongly indicate chemical action. This force now operates by three methods, as already described in the three sections on the three "assumptions:" first, by elements that afford vastly more heat than those in our earth; secondly, by probable conditions for producing heat different from any thing in our small experiments; thirdly, by the formation of new fuel, either as new combustible compounds, or as new simple elements from previous incombustible material. Materials may be incombustible in one state, but very combustible in another allotropic state. Carbon, in the form of plumbago, is so incombustible as to be used as crucibles in the hottest furnaces; but in different states it is the best of fuel. To give a full view of this theory before the discussion of the nebular theory, is inconvenient. It would occupy too much space, and the reader is not prepared. Hence, a further exposition of the theory of chemical force will be deferred to the section reviewing other modern theories of creation.

PART III.

THE ORIGIN OF THE STARS.

SECTION XVI.

THE UNIVERSAL DIFFUSION OF MATTER.

As the round and beautiful rain-drops which shine in the rainbow derive their origin from a condensation of the moisture in the air, so the round and beautiful stars which shine in the sky derived their origin from a condensation of matter formerly diffused through all known space. But what held matter in this diffused condition? What caused it to condense? And what is the force which gave their motions to the stars, and arranged them in their curious clusters? The solar system, for instance, is one of the smallest of these clusters; it consists of more than a hundred members. They are woven in a web; or rather, they dance around in complicated circles, each one having two motions, and some of them three. The moons rotate on their axes, and they revolve around the planets, while, at the same time, they are revolving around the sun. The forces which arranged the stars into systems, which gave them their astonishing velocities, which caused their condensation from a nebulous condition, and which formerly held all matter suspended in that nebulous condition, we are now about to explain. We are entering upon a grand and sublime inquiry. It is not simply the creation of the world, about which Milton sang; but the creation of untold millions of suns and of worlds. His lines, in beginning his great poem, are appropriate for us now:

> "Thou, O Spirit, that dost prefer
> Before all temples the upright heart and pure,
> Instruct me, for Thou knowest.
> What in me is dark,
> Illumine; what is low, raise and support;
> That to the height of this great argument
> I may assert Eternal Providence,
> And justify the ways of God to man."

But why do we assume that all matter was formerly diffused equally, or nearly equally, through all known space? Because from this assumption, and from the action of well-known forces, we can demonstrate with mathematical certainty, that just such clusters of stars as our solar system, with precisely such velocities, must result. The aim of our argument will be to show, that the production of our solar system, with its strange peculiarities, is one of the inevitable consequences of such an assumption. In all this we may recognize a wise, a mighty, and a benevolent First Cause. Just as He now creates every apple on the trees by the agency of definite forces, so He has created the stars. Our duty is to study the action of these forces, both in the formation of an apple and of a world.

By what force, then, could matter be diffused through all space? By the force of atomic repulsion. By this force, an ounce of hydrogen is now diffused through a quarter of a million times more space than an ounce of platinum; that is, by this force its atoms are held a quarter of a million times further apart. It does this against an atmospheric pressure of one ton on every square foot of space; and how far it can possibly hold the atoms of matter asunder without such pressure no one can tell.

By what force could this repulsion be overcome, and condensation be carried on? By the chemical force; the same force that overcomes the repulsion in both hydrogen and oxygen, and causes their condensation into water. The most mysterious of all known truths is that of the conversion of forces. Mechanical force, such as the blow of a hammer, may be converted into the force of heat, and this heat into the force of electricity, and this electricity into the chemical force; all these forces successively being in precise and definite proportions to one another. But how the conversions take place, we have not the least idea. We may well assume that the repulsive force which held all matter diffused through space, was converted into chemical force. But, as in the other cases just mentioned, we cannot tell how. Neither can we tell what should cause chemical action to begin, any more than we can tell what causes chemical action now. We simply know its possibility now, and such possibility may have been then. We may suppose that even then, by some means unknown as yet, matter was modified into two or more different elements, thus allowing their chemical combination by different degrees into various compounds. But we know that the necessary consequence of chemical combination is condensation. Therefore, after chemical action began, matter immediately occupied less and less space. It must, by the same process, continue indefinitely to occupy a smaller amount of space. Hence, after an inconceivable number of millions of years, matter, from being universally diffused, is now contracted into solid and liquid globes, like the sun and the earth. The chemical action now going on in the sun, we may regard as the continuation of the long process; and the force which now proceeds from the sun in the form of light and heat, we may regard as but the conversion of the repulsive force which once held matter in a state of universal diffusion.

Thus when we assume that all matter was diffused through all known space, we assume no known impossibility; for we know a force whose office is diffusion: and when we assume that a gradual contraction has been going on, we assume no known impossibility; for we know a force whose continued action is followed by a continued contraction.

We cannot suppose that this primitive nebulous matter was luminous. Even the sun when expanded to the orbit of Mercury, was thirty times more rare than hydrogen; and rare gases, such as hydrogen and oxygen, when entering into combination, produce no light, or the feeblest of all lights. Hence we cannot believe that we now see patches of luminous nebulous matter in the far distant space condensing into stars. Those apparent nebulæ must undoubtedly be systems of stars too distant to be separately distinguished.

We cannot suppose that matter was originally diffused through space by the repulsion of heat; because heat is a well-known effect produced by some action. We can conceive of no action to produce such an enormous amount of heat. And if all space were pervaded by such heat, it must have remained. It could not have radiated away: for where could it go?

But with the causes of this primitive diffusion and contraction we have properly in the remaining part of this volume nothing to do. The causes above mentioned are merely suggested, to show that our assumption of a primitive diffusion and a gradual contraction, is not absurd or unnatural. It is a fair assumption, and in harmony with scientific principles. Our real business henceforth in this volume is, simply with contracting nebulous matter, from whatever cause; and I am to show that gravity, acting on a contracting nebulous mass, produced all the wonderful arrangements of the solar system. Gravity separated that mass into more than a hundred large bodies, and gave them their forms and their positions. Gravity imparted their swift velocities in all their complicated motions, and directed the shapes and the positions of their orbits. All this is to be shown, not by loose conjectures, but by mathematical demonstration, and by strict reasoning from well-known dynamical laws. In a word, I am to prove that the origin of the stars has been from a gradual condensation of nebulous matter, and that the cause of their motions was gravity. The course of the proof will be by showing that our solar system, in all its peculiarities, is the inevitable and necessary result of gravity acting on such a mass. The only exceptions relate to the densities and sizes of these great bodies, which, like the light and heat of the stars, must be looked upon as caused by chemical action.

SECTION XVII.

THE NECESSITY OF ROTATION IN A CONTRACTING NEBULOUS MASS.

IF, from a principle of repulsion, all matter were diffused through all space, and if from any cause a gradual contraction were to ensue, then that matter would separate into huge masses, like clouds, when the generally diffused moisture of the atmosphere is clearing away. But these huge masses would not be round. That would be impossible, because round bodies cannot fill space. They would naturally be angular, and more or less irregular, like compact cumulus clouds. But gravity would soon begin to make them round, as it now keeps them round

This it would do, not by bringing down their prominences straight toward the centre, for the principle of repulsion would prevent that. These prominences, angular projections, would slide down laterally into the adjacent depressions, in every direction. But in so doing many currents would be formed, and these currents, by the composition of forces, would all result into one. Imagine our own globe to be all water, and cubical or nearly cubical in form. Then, if left to itself, it would soon be round, by the action of gravity. But what mighty currents would be put in motion in assuming this rotundity of form! The slightest imaginable irregularity in the cubical shape would produce irregular and unequal currents, and these would oppose and act on one another, until at last only one current would result. We may convince ourselves of this by rapidly drawing our hand in various directions through a basin of water. However many currents we make, they will soon counteract one another, and flow into one. Even if the nebulous globes had been miraculously created round, then by their motions and their mutual perturbations, their roundness would have been lost, and in recovering their rotundity of form again by the action of gravity, currents must have originated at their surfaces.

But, in the nebular globes, after a single circulating current had been established on their surfaces, that current could not stop as long as contraction in their bulk continued. It must keep on, and be continually hastened in its velocity. This is a fundamental truth in the reasonings in this volume. It accounts not only for the rotations of the countless globes that roll everywhere through space, but also for the velocities of the planets and satellites in their orbits, and for the proper motions of the fixed stars. We must therefore pause here, and have this great truth made evident and clear beyond all doubt.

After the many currents at the surface of a globe had subsided into one, that one would constitute the rotation of the globe. It would now have its axis, its poles, and its equator. Think of the separate particles of matter around the equatorial zone. As the globe contracts, the course of every one of these could not be tangential; or, in other words, could not be at right angles to the radius of the globe. They must all slope a little toward the centre of the globe, simply because the globe contracts. Going round, and at each instant gradually approaching toward the centre, the path of every particle must be a spiral, and an inclined plane: a plane spirally inclined from the original position of the surface, round and round, no matter how often, until it ends as near the centre of the globe as the surface approaches that centre by contraction. Reflect now on the consequence of this descent down an inclined plane. Gravity is all the while acting on every particle as it descends, and hastening its speed. Hence the rotating surface, as the globe continues to contract, will rotate all the time more and more rapidly. We may say that as every particle marches onward, the ground at every succeeding step sinks down, and renders the entire course descending.

FIRST ILLUSTRATION.—In the figure, a sector of a circle,
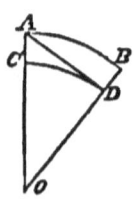
A and B are equally distant from the centre O, and so are C and D. The particle in motion around the centre does not proceed from A to B, but in consequence of the contraction of the globe it proceeds from A to D; that is, its track is always inclining toward the centre, and gravity huries it on.

SECOND ILLUSTRATION.—Many years ago, when the Long Island Railroad was first built from Brooklyn, opposite New York, to Jamaica, a distance of about twelve miles, one evening a few boys tried their strength to move an empty car standing on the track. They at first could move it very slowly, and finally as fast as they could walk; but when they stopped the car did not stop. The road seemed perfectly level, but it had an imperceptible inclination all the way, and as the track happened to be quite clear, the car regularly increased its speed, and soon the astonished people of Brooklyn saw it come thundering along until it leaped off the wharf, and plunged foaming into the harbor. If the road had not terminated there, but had continued in a great circle round the globe, the car continually increasing its speed, would have returned to the locality of Jamaica again, but down several thousand feet below. This car may be taken to represent a particle at first only slightly in motion on the surface of a contracting fluid globe. The courses of the two would be precisely the same. The continued and regular contraction of the globe would be equivalent to the inclination of the road. Gravity, alike in both cases, is the motive power.

THIRD ILLUSTRATION.—Water descending through a funnel is another illustration. If any portion be slightly moved at right angles to the radius of the earth, then by the gradual settling of the mass in the funnel, the moved portion of the water will run down a spirally inclined plane. It will continue to increase its rotating speed, hindered only by friction. Because it moves nearly horizontally, and settles at the same time toward the centre of the earth, therefore it moves on an inclined plane. There is no difference in principle between this and the car on the inclined railroad. Here the sides of the funnel prevent the water from taking a straight course. It is as if the Jamaica road had taken a spiral course, round and round, either to the right hand or to the left, until it had reached below that town a few thousand feet. A particle of water in the emptying funnel, and the car on the inclined road, are both alike moved by the force of gravity down an inclined plane. Both are like a particle on the surface of a contracting nebulous globe. If only a single particle moves, it will increase its speed, and move a second and a third by contact, and so on until all the particles move and the globe rotates.

FOURTH ILLUSTRATION.—An eagle poised in the air moves forward with an arrow-like velocity, without once flapping its wings. It merely balances itself so as to move down an easily inclined plane. It is the image of a particle of matter on the surface of a contracting nebulous globe. After it has begun to move horizontally, the contraction of the globe forms the inclination of the plane.

So many illustrations of one truth may appear superfluous; but they are not. Already some intelligent men have complained of their number as needless, and afterwards they have expressed a doubt of the certainty of all these reasonings. That doubt arises from their not having firmly fixed in their minds this fundamental truth, that, on the surface of a contracting nebulous globe, if motion once began, IT COULD NOT STOP! ITS VELOCITY MUST INCREASE! Only one ounce of matter in motion would soon move more rapidly, because gravity would take it down hill. That ounce, by friction, would move other ounces, and so on, until THE WHOLE GLOBE ROTATED! But all the great globes of space rotate. The satellites, the planets, the sun, which is a fixed star, and also the other fixed stars, rotate, as we know by the periodic variations of their light. This rotation in all the stars we now see to be the inevitable consequence of gravity acting on a nebulous mass.

But rotation produces oblateness. All these great stars, as far as can be ascertained, are more or less oblate. Therefore, the oblateness of the Earth, of Jupiter, of Saturn, and of other orbs, is another inevitable consequence of gravity acting on a shrinking nebulous mass.

But it may be said that oblateness is not produced by gravity, which is the centripetal force, but by inertia, which is the centrifugal force. This has hitherto been the general scientific opinion, but now we have new and more enlarged views. Faraday says, with beautiful simplicity, that "inertia is always a pure case of the conservation of force." A ball in motion is endowed with a certain amount of force, the force of inertia. But this is precisely the force given by some anterior impulse. This anterior impulse is conserved or converted in the form of inertia. So in a rotating globe; the more rapidly it rotates, the greater is the centrifugal force, or the force of inertia. Therefore its inertia arises from the rotation. But the anterior impulse which caused the rotation is gravity. Therefore gravity is conserved, or converted, in the form of inertia, and made to act antagonistic to itself. Gravity is not only the centripetal force, but it also gives origin to the centrifugal force. In all other departments of astronomy, even in the revolution of one orb around another, we will soon see that the centrifugal force always had its origin in gravity—the same as heat may arise from a blow of a hammer, or as, we suppose, chemical force may arise from the atomic repulsive force which once expanded matter through all space, and which now appears as light and heat from the sun and fixed stars.

Four classes of facts—the round form, rotation, oblateness, and centrifugal force among the stars—are now proved to be the inevitable consequences of gravity acting on a contracting nebulous mass. Such facts will soon be found so numerous as to leave no doubt about the origin of the stars, or the cause of their motions.

SECTION XVIII.

THE VELOCITY OF ROTATION.

Having proved the necessity of rotation, the next procedure in order is to demonstrate the velocity of that rotation. It has been shown that the path of every particle on the surface of a contracting and rotating nebulous globe inclines toward the centre of that globe. As it goes round and round, it constantly approaches nearer and nearer to that centre. The particles therefore move down an inclined plane, and become subject to the law of motion down such a plane: one of those laws is very simple, and thus expressed: A body moving down an inclined plane acquires the same velocity as when it falls the height of the plane. A ball rolling from A to B acquires the same velocity as when falling from A to C. Both in rolling down the plane and falling perpendicularly, the velocity is continually increasing, and it increases by the same law in both cases. When the ball has rolled to the points d, e, f, g, h, it has the same velocity as when it has fallen to the corresponding points d', e', f', g', h'. The times, however, are very different, depending on the inclination of the plane. When the plane is inclined infinitely little, the motion down the plane will require a time infinitely longer, although the ultimate velocities will be identical.

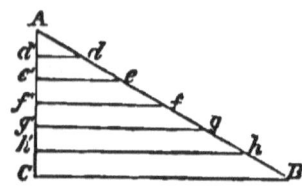

But it may be said that a spiral line, the path made by a particle on the surface of a contracting and rotating nebulous globe, is not an inclined plane, and therefore cannot fulfil the law of motion down an inclined plane. To this it must be answered, that a true inclined plane seems perfectly straight; but when true theoretically, it must really not be a plane, but a curve, inclining in every part of its length with the same angle to the surface of the earth. As the surface of the earth is not a straight line, but a curve, therefore the inclined plane, to preserve the same angle of inclination, must curve correspondingly. To be perfectly true and constant in its angle of inclination, a plane must in reality not be a plane. It must be that kind of a curve called a spiral. It must resemble a segment in the mainspring of a watch partly uncoiled; one end being at or near the centre, and the other at the circumference of a circle. This is the theoretical idea of a perfectly true inclined plane, although the terms are paradoxical.

The annexed figure represents the inclined plane, down which descends every particle of matter in a rotating and contracting nebulous globe. The spiral line A B represents the length of the plane, and the straight line A B represents the height of the plane. A particle of matter moved by gravity and without any friction, will gain the same velocity in passing down the spiral as down the straight line. The times of the two passages, however, will be vastly different, depending on the number of coils.

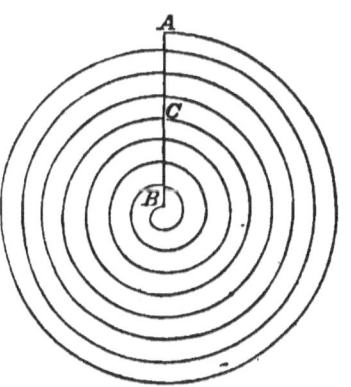

Hence, if our sun were ever expanded to a point far beyond the orbit of Neptune, then in contracting it must have rotated, and every particle of matter on the surface of the equatorial zone must have moved down a spiral plane coiling many million times within itself, and the velocity which those particles acquired at the orbit of every planet must have been the same, friction excepted, as if they had fallen to the orbits of those planets in a straight line from their starting-points toward the centre of the sun. Therefore, in order to determine the velocity of rotation of any contracting nebulous globe, at any point in the course of its contraction, we may calculate the velocity of a fall from the starting-point of contraction. Thus in the figure the velocity of rotation, when the surface by contraction had arrived at C, would be equal to that of a fall from A to C.

Proceeding on this undoubted preliminary truth—the law of the inclined plane—my course will now be to prove that when our sun was in a nebulous condition, the power of gravity must have given it a velocity of rotation at its equatorial zone, when at the orbits of the several planets, just equal to the present velocities of these planets, especially when taking into account a necessary retardation by friction, which was almost infinitesimally small. I will next show the absolute necessity and the amount of this retardation by friction. And then I will show that gravity separated the several planets from that equatorial zone of the sun under such velocities, and moulded them in their present spheroidal forms.

First, then, I am to show the velocities of rotation of the sun at the orbits of the several planets; or, what is the same, the velocities of the falls toward the sun at these orbits. The velocities of these falls must depend on two things; the distances of the falls, and the amount of the force of gravity in each one. Gravity varies its power. It decreases at different distances from the sun by a well-known law; the amount of force being inversely as the squares of the distances. A mass of matter which would be attracted with the force or weight of 409,000,000 pounds at the sun's surface, would be attracted with the force of 902 pounds at

THE VELOCITY OF ROTATION. 229

the earth's orbit, and of only one pound at the orbit of Neptune. It would be attracted with only the one hundredth part of an ounce at forty times the distance of Neptune. Hence, each of these falls to the orbits of the several planets has not only different distances but different motive powers. Hence also we need not begin our calculations at any definite point far out in space where we may suppose the falls began, because gravity was so exceedingly feeble at such distances as forty times, or a thousand times the distance of Neptune. We may practically omit such small quantities, and assume at Neptune's orbit all the velocity which a body possibly could acquire, when it has arrived at that point from infinite distance in its fall toward the sun.

The general reader may not wish to go through the details of the mathematical calculations by which are obtained the velocities of the falls to the orbits of the several planets. They are therefore put in the Appendix I. at the end of this volume, and special pains have been taken to render them so plain as to be easily understood by persons of very moderate mathematical acquirements. And now the momentous question arises, with all the interest that can be awakened by the creation of the universe, what must be the velocities of these falls to the orbits of the several planets? Is the force of gravity toward our sun, far away beyond the solar system, sufficiently powerful to give a velocity to a falling body equal to the velocity of Neptune, when it would arrive at the orbit of Neptune? Would its velocity be equal to that of our earth, 68,000 miles per hour, at the earth's orbit? Would it equal Mercury's velocity, 110,000 miles per hour, at that planet's orbit? These are great questions, and after due carefulness in computation the answers are found that the velocities of these falls precisely equal the velocities of the several planets, when just allowance is made for a small retardation by friction, which the equatorial zone of the sun suffered in its descent down the inclined plane. As the planets were formed from that zone, they had the same amount of retardation. An exhibition of the results of this computation is given in the following table. The first column shows the calculated velocities of the planets; the second column shows the subtraction to be taken from the first column on account of retardation by friction; this subtraction, when performed, gives the third column, which coincides precisely with the present actual velocities of the planets. All these velocities are expressed in miles and decimals of a mile per second. Thus the velocity of Uranus is four miles and three hundred and sixty-nine thousandths of a mile per second. The fourth column shows the retardation in the form of percentage:

	Calculated Velocity.	Retardation.	Actual Velocity.	Retardation estimated by Percentage.
Neptune	4.937	1.44	3.491	30
Uranus	4.735	.365	4.369	8
Saturn	6.958	.762	6.196	11
Jupiter	9.305	.916	8.389	10
Hygeia	11.52	.97	10.55	10
Astræa	12.41	1.05	11.36	8
Mars	16.44	.94	15.50	6
Earth	20.72	1.59	19.13	8
Venus	23.83	1.33	22.50	5.6
Mercury	33.20	2.44	30.76	7.4

Before proceeding to the exposition of the cause and the mode of ascertaining the amount, of retardation by friction, it is proper here, in presence of the above table, to say, that where the nebulous sun was most dense, between the orbits of Venus and Mercury, the retardation was only one per cent. through a contraction of 9,000,000 in diameter. Between the orbits of Neptune and Uranus, where the nebulous sun was most rare, the retardation was only one per cent. through a contraction of 250,000,000 miles in diameter! That is, in proceeding down an inclined plane 250,000,000 miles *in height*, the loss by friction was only one per cent. So very near is the coincidence between the present actual velocities of the planets and their velocities due by the force of gravity!

A clearer idea of the small loss of velocity by friction is obtained not by viewing the heights but the lengths of the planes. Their lengths were to their heights doubtless as many thousands to one; but assuming the difference as only one hundred to one, then the loss by friction of the velocity due to gravity must have been, between the orbits of Venus and Mercury, one per cent. down a plane 450,000,000 miles in length; and, between the orbits of Neptune and Uranus, one per cent. down a plane 12,500,000,000 miles in length.

SECTION XIX.

THE RETARDATION FROM FRICTION.—SURFACE ROTATION.

I PROMISED in the last section to show the absolute necessity and the amount of retardation by friction. This can now easily and clearly be done.

A nebulous globe perfectly round and perfectly homogeneous, and undisturbed from without, could not acquire a rotary motion by contraction. However slowly or however rapidly it might contract, each particle would move in a radial line toward the centre. We can think of nothing to turn any particle out of such a line. But if a globe were not perfectly homogeneous, or if its rotundity were disturbed by perturbations from without, then the case would be different. Then the particles in the process of contraction would be turned away from the radial lines; and having once begun an oblique motion toward the centre, their rotating course would begin and accelerate.

But the shapes of the primitive bodies, the nebulous masses in the beginning, could not have been round; for, as already mentioned, round bodies cannot possibly fill space. In harmony with the laws that govern matter, the separate nebulous masses must have been irregular in shape, some more and some less. Gravity would soon operate to make them round. It would be on their surfaces that this operation for producing rotundity would take place. The prominences at the surface would be prevented by atomic repulsion from taking the direct course to the centre in obedience to gravity. Therefore those prominences would slide down laterally, by the force of gravity, into the neighboring depressions. Every such sliding down would produce a current. The number of these several currents would at least equal the number of prominences, and these would be numerous. But in a fluid of definite size and shape, all these currents would act and react against one another, and by the composition of forces they would result into one. As these several currents would be originated only to level down the prominences and to fill up the depressions on the surface, so they would be confined to the surface. Their depths would depend on the heights of the elevations and the depths of the depressions. When all the currents had coalesced into one, the depth of that one could not considerably exceed that of the several primitive currents. This single resulting current would constitute the rotation of the new-born globe, and the rotation would be only on the surface. For in all this we see nothing to disturb the interior, or to turn away its particles from their motion in radial lines, until they are moved by friction with the rotating surface. That friction would retard the surface. The amount of this retardation would depend on the quantity of matter in the moving surface, compared with the quantity of matter to be moved by that surface. If the mass of the moving surface were very light compared with the mass of the inert interior, then the retardation would be very great.

In a contracting nebulous globe we know of no reason why any one portion should contract more rapidly than another. Every part we must suppose to contract equally; and hence while the surface is contracting and settling toward the interior, that interior is contracting and settling in radial lines toward the centre. In the beginning of this process of contraction, therefore, the rotating surface would come comparatively little in contact with the great mass of the interior, because that interior would be settling away from it, and condensing around the centre. Hence in the beginning, when the nebulous matter was more rare, the retardation from friction would not be so great as when toward the last the rotating surface came in contact with the whole of the condensed interior in the neighborhood of the centre. Hence in our solar system, we must *à priori* look for a greater and greater amount of retardation at each succeeding planet, beginning with the outermost; and on the present surface of the sun we must look for by far the greatest retardation of all. The orbit of Mercury is 66,000,000 miles in diameter, and within that wide space the largest portion of the matter of the sun had settled before its rotation began, and therefore the retardation of the surface of the sun as it now rotates, is enormous compared with the retardation of the planets. Let us now examine the retardation of the several planets, and see whether they follow the anticipated law of being more and more retarded as they are nearer and nearer to the sun:

	Percentage of Retardation from former table.	Interplanetary spaces in miles.	Miles of contraction in radius for a retardation of one per cent.	Comparative amounts of retardation.
(1)	(2)	(3)	(4)	(5)
Neptune......	30			
Uranus.......	8	1,000,000,000	125,000,000	1
Saturn.......	11	887,500,000	80,000,000	1¼
Jupiter.......	10	398,900,000	38,000,000	3
Mars.........	24.5	338,500,000	13,810,000	9
Earth........	8.3	48,000,000	5,783,000	21
Venus	5.6	25,500,000	4,554,000	27
Mercury......	7.4	33,000,000	4,459,000	28

Column two, taken from the former table, represents the percentages of retardation in all the planets. The asteroids are omitted, for obvious reasons; and the large percentage opposite Mars represents all the retardations of all the asteroids as well as that of Mars; that is, all the retardation that occurred in the contraction from the orbit of Jupiter to that of Mars. The large percentage opposite Neptune harmonizes beautifully with the whole theory of retardation, as will soon be pointed out; but, in the mean time, the percentage of this planet will not be included in the general remarks now to be made. With these qualifications, and especially when looking at a preceding table containing two of the asteroids, it is noteworthy that the retardation numbers in column two are confined within narrow limits, ranging from 11 for Saturn down to 5.6 for Venus. But in this column of percentages, at first view we see no indications that the law of retardation is a regular increase as anticipated *à priori*, from the outermost to the innermost planet. On the contrary, the reverse seems rather the case. This, however, is only from a first and superficial inspection. The real amount of retardation suffered by any planet must be estimated by the distance through which contraction has been going on since the formation of the next outer planet, or, in other words, by the distance it has travelled down its inclined plane. If the equatorial zone of the nebulous sun has been retarded ten per cent. in a contraction of 1,000,000,000 of miles, this is equal to a retardation of one per cent. in a contraction of 100,000,000 of miles. This idea must be clearly understood before proceeding further. We will then see that the outer planets with the same percentage numbers, have really suffered less retardation than the inner planets, because the retardations of these outer planets were taking place through longer distances of travel down their inclined planes. Column three represents the interplanetary spaces, their distances from the sun being calculated on the basis of 92,000,000 miles for the earth's distance. These interplanetary spaces represent the distances of contraction through which the retardations of the several planets were accumulated. For instance, during the contraction of the nebulous sun of 1,000,000,000 miles in radius, the velocity of Uranus was retarded eight per cent. ; and during a contraction of 25,500,-000 miles, Venus was retarded 5.6 per cent. In these two cases Venus suffered a retardation twenty-seven times the greater. To find the relative amounts of retardation, therefore, we must divide the interplanetary space outside of each planet by its own percentage of retardation : in this way we obtain the column four, and this shows how many miles the original nebulous sun contracted in radius in suffering a retardation on its equatorial zone of one per cent. Here we arrive at our desired result ; for this column four shows the increase in amount of retardation from one planet to another, beginning with Uranus and ending with Mercury. We see how it agrees with our *à priori* anticipations. During the contraction from Neptune to Uranus, the retardation averages only one per cent. for every 125,000,000 miles in *radius;* and during the contraction from Venus to

Mercury, the retardation averages one per cent. for every 4,459,000 miles in radius. This latter is the greatest of all the retardations; nevertheless, in *diameter* of contraction it is only one per cent. for every 9,000,000 miles! This is truly an insignificantly small retardation, and warrants us in saying that the coincidence between the actual and the calculated velocities of the planets proves gravity to be the cause of their velocities. Between Neptune and Uranus the retardation was only one per cent. on an average, for every 250,000,000 miles of contraction in diameter! And the regular increase in amount of this retardation from planet to planet, according to *à priori* expectation, adds still greater firmness to our conclusion respecting gravity as the cause of these several planetary velocities.

The comparative amounts of retardation suffered by each planet are best shown by column five. The numbers in that column are obtained by dividing the number in column four opposite Uranus, by the numbers in the same column opposite the other planets. These quotients show how many times greater the retardation was in those planets than in Uranus. Thus the retardation between Jupiter and Saturn averaged three times as much as that between Uranus and Neptune. The small fractions have been omitted in column five, as is often done in astronomy, and, with these omissions, it is remarkable that the numbers have all a simple relation to three, except the last opposite Mercury. Again, it is remarkable, that if we multiply the number opposite Uranus by three, we obtain the number for Jupiter, passing over Saturn; and in multiplying the number for Jupiter by three, we obtain the number for Mars, passing over the asteroids; and in multiplying the number for Mars by three, we obtain the number for Venus, passing over the Earth. There must have been a physical cause for this curious progression in the rate of retardation. It must have been caused by the increasing rate in which the rare and light rotating exterior came in contact with the dense, inert, unrotating interior, and thus the friction was augmented.

But it may be objected, that we are unable to form an estimate of what the friction should really be in the nebulous masses; that therefore the loss of velocity by friction cannot be calculated by itself independently; and that in these reasonings its amount is merely inferred from the difference between the actual and the calculated velocities of the planets. To this it may be answered: 1st. This is true of friction in all other cases, even in ordinary mechanics. Its amount is always inferred from a comparison between a force and its net effects, precisely as here. 2d. Other disturbing forces are taken into account in astronomy, and due allowances made for them, even when they are not numerically calculable. For instance, the satellites of Jupiter do not fulfil Kepler's third law. The squares of their times, divided by the cubes of their distances, should give the same quotient for all these satellites. But the actual quotients are 8.6, 8.2, 8.1, 8.1. Here is a discrepancy of six per cent., or .06. This, however, does not shake our faith in Kepler's third law, because the discrepancy is plainly due to the oblateness of Jupiter, although the influence of that oblateness has never been calculated. Sometimes in astronomy there are discrepancies whose causes are not only not calculable, but even not known; still they have no influence in preventing belief. At all the outer planets, beginning with Jupiter, there was, until two years ago, a considerable difference between the calculated amounts of the centripetal and centrifugal forces. That has just now been removed by the new determination of the earth's distance from the sun; making the earth 3,000,000 nearer, and Neptune 90,000,000 miles nearer the sun. From the greatness of this error every one may form a judgment of the considerable discrepancy formerly between those two forces at the outer planets; especially since a change of distance increases the one force and diminishes the other, by the square of that distance. But it had no influence toward destroying belief in gravity as the centripetal force, for astronomers were sure there must be an error somewhere. Such discrepancies are gross compared with the infinitesimally small one in the case of Uranus, whose loss of velocity by friction is only the $\frac{1}{250000000}$ of one per cent., in a contraction of the nebulous sun of one mile in diameter. Small as this discrepancy is, we can show its cause. Therefore I do not know why we should hesitate to announce that there is a perfect coincidence between the actual velocities of the planets and their calculated velocities as produced by gravity.

We are now prepared to understand the apparently large percentage opposite Neptune in the table. That large number speaks not of the greatness of retardation out there, but of the vast distance through which the nebulous sun contracted to produce that amount. When we come to calculate the sizes of the planets in their former nebulous condition, we will see that Jupiter contracted one hundred times the distance between his centre and his outer satellite, before producing that satellite. If the sun contracted proportionally far before producing Neptune, then the rate of retardation experienced by Neptune must have been one per cent. for many thousand millions of miles; that is, exceedingly less than was suffered by any other planet.

This and the previous section is further illustrated in Appendix II.

SECTION XX.

FROM A ROTATING NEBULOUS SUN GRAVITY SEPARATES A CLUSTER OF PLANETS AND SATELLITES.

HAVING shown how gravity causes rotation in a contracting nebulous mass; having given the process to calculate, from time to time, the several increasing velocities of that rotation; and having pointed out how that rotation becomes retarded by a necessary friction, we next proceed to prove that a nebulous sun, by the continued action of gravity, may be separated into many smaller parts, and may thus be surrounded by a halo of a hundred attendants.

I have already proved that oblateness in a contracting nebulous globe is caused by inertia, and that this inertia is a clear case of the conservation of the force of gravity. The longer gravity applies itself to any particle of matter free to move, the more will its own force be converted into the force of inertia. Therefore, whatever is done by inertia—the centrifugal force—is done by gravity conserved.

The strength of this centrifugal force in any rotating globe, and at any given distance from the centre, depends on the velocity of rotation. It is a matter of easy mathematical demonstration to show that by a fall, or by a descent down an inclined plane, a velocity of rotation may be imparted by gravity more than sufficient to produce a centrifugal equal to the centripetal force. But by a natural retardation from friction, a velocity of rotation is secured at the several planetary orbits, no more than sufficient to equalize the centrifugal and centripetal forces. In consequence of this equalization of these two forces, the matter on the equatorial zone of the nebulous sun would be held in equal balance. It could neither approach toward the centre, nor fly off in a tangent from that centre. It must continue the same distance from the centre, and revolve in an eternal orbit. As the globe continued its contraction, it would, therefore, separate from the matter on its outer equatorial zone, and that zone would circulate with its accustomed velocity as a separate body, a ring around the great central mass.

It is important to observe that such a ring would separate only from the equator of the nebulous globe—a narrow belt, say of five degrees on each side of the equator. This would happen because that zone is farther from the axis of rotation, and therefore, compared with all other latitudes, its centrifugal force would be the greatest, and its centripetal the least. While these two forces would be equal at the equator, they could not be equal in any other part of the globe. In all other latitudes the centripetal force would be the stronger of the two, and no material could separate. The distance of the equatorial zone from the axis of rotation, when compared with all other latitudes, would be especially great on account of the extreme oblateness of the swiftly rotating nebulous globe. From the extreme rarity of the sun when expanded to the orbits of the planets, and from its rapid rotation, the oblateness was such that the equatorial was about double the polar diameter. This oblateness must have given a sharp edge to the globe at the equator. It is easy to see how that sharp edge—a very narrow belt on the equator—must have first been balanced between the two forces, the centripetal and the centrifugal, and rendered stationary, neither flying off like a stone from a sling, nor approaching any nearer to the centre.

When any portion at the equator was separated as a distinct ring in consequence of the departure by contraction of the central globe, then other portions must soon have occupied its place, always completing the same oblateness, and the same sharp edge on the equator. Then another ring would be formed, and in this manner successive rings would be separated.

Let us now trace the future history of these rings. Far away at the orbit of Neptune, the force of gravity toward the sun is very feeble. There is also at the same point a force of gravity, though still feebler by far, toward certain fixed stars. This latter force would act on the ring, the same as Neptune acted on Uranus, and led to his own discovery. This action toward the fixed stars, being greater in some parts of the ring than in others, would break the ring. Moreover, as we suppose chemical action was boiling and raging in the great nebulous mass, the same as it now renders the sun the most unquiet of all known places, we cannot conceive how the ring could be perfectly even in form. Its thicker portion, by the force of gravity, would cause a fissure in the ring. When, from either or from both of these causes, a fissure had occurred, then, from the continued action of gravity, that fissure would widen. It would continue to widen until at length all the matter of the ring would be collected in one nebulous mass, and this new mass would become round and rotate from the same force which rounded and rotated the original mass. Here would revolve a rotating nebulous planet around the central nebulous sun, and this planet would by gravity break the next interior ring. Here, again, would revolve and rotate a second planet; and this would in turn break the next interior ring; and so the process would go on, until all the rings were broken, and as many planets, with the same velocities as the rings, would revolve around the central sun. Here we behold the origin of the planets and of planetary motion. Evidently the cause of their motion was gravity, and they were separated from the central sun by gravity conserved as inertia or centrifugal force. By gravity also they were rounded, rotated, and again, by gravity conserved, they were made oblate.

The planetary orbits are nearly circles, because the materials of the planets moved in circles when they formed part of the equatorial zone of the sun.

There is an exact balance between the centripetal and centrifugal forces; and this was adjusted by the gradual acceleration of the materials of the rings down the inclined planes. When the velocity of the matter on the equatorial zone with its centrifugal force became great enough to equal gravity, then that matter ceased approaching the centre, and remained poised in equal balance.

The planets all move nearly in the same plane and in the same direction, because they were derived from the same globe, which always rotated in the same direction, and with its equator in nearly the same plane.

The planets now move in the same direction as the sun rotates, because there has been no change in either the sun or the planets; gravity having originally impelled the materials of them all in the same direction when they were an undivided mass.

The planets move nearly in the plane of the sun's equator, because the sun's equator has remained nearly in the same position. They have very different velocities, because they were separated from the sun's equator when, at their several orbits, that equator had very different velocities.

In a former section we saw how rotation, oblateness, rotundity of form, and centrifugal force as it exists among the stars, are inevitable consequences of gravity acting on a contracting nebulous mass. We have now seen other inevitable consequences of such action. We have seen that at first the rotation of a nebulous globe is wholly on the surface; that the velocity of this surface rotation must be a little retarded by friction on the unrotating interior;

THE ORIGIN OF THE SATELLITES. 243

that in a nebulous globe whose mass is equal to that of our sun, the force of gravity is powerful enough to impart a velocity of rotation equal to that of the several planets when its surface was at their orbits, and also enough in addition to overcome the small amount of necessary friction. The inevitable consequence of these velocities of rotation was to separate equatorial rings whose centripetal and centrifugal forces were equal, and which therefore could approach no nearer the centre of the sun. These rings by the force of gravity must break and be formed into planets having the same velocities, the same orbits, and the same directions of motion which the planets now have. All these are necessary consequences of gravity acting on a condensing nebulous mass, and they show with absolute certainty that gravity was the cause of the motions of the planets in their orbits. Gravity is the agency in the hand of Omnipotence that sent our earth through space with a velocity of 68,000 miles an hour.

SECTION XXI.

THE SATELLITES.

WE have seen how a great contracting nebulous mass must become a globe, must rotate, and produce rings. We have seen how these rings must break, must contract into rotating planetary globes, and revolve around the central globe. For the same reasons these planetary globes, if large enough, must also produce rings; and these secondary rings must break, contract into globes, and revolve as satellites around the planets, thus forming planetary systems. We now apply the inclined plane theorem to these planetary systems. We may determine how large were these nebulous planetary globes. We may see whether they were large enough to produce rings; and if so, we may decide exactly what would be the velocites of the rings, and consequently the velocities of their resulting satellites around the planets. Would their velocities be equal to the actual velocities of the many satellites? For instance, could the velocity of the inner satellite of Jupiter, 38,000 miles per hour, be produced by the gravity of Jupiter in bringing the matter of that satellite down an inclined plane? What must have been the size of the nebulous mass whose condensation formed our earth? Did it extend very far beyond the orbit of the moon, and could its equatorial zone, in the course of contraction and rotation, acquire the velocity of the moon when it had contracted as far as the orbit of the moon?

The sizes of the planets when in their nebulous conditions may be determined in this way: We may calculate how dense the sun must have been when it was expanded as far as the orbits of the several planets. Then it is plain that at these several distances the rings forming the planets must have been many hundred times less dense than the average densities of the sun; the same as the atmosphere of our earth is much less dense up at the surface than down here. But after all their contraction during their condition as rings, and during their aggregation into globes, we may assume as a moderate estimate that when their rotation began, they were fifteen times less dense than the average density of the sun when expanded to their orbits. As the next step, we must take the masses of the several planets and calculate HOW LARGE each one must have been when expanded to these densities; that is, fifteen times less dense than the average of the sun when enlarged to their orbits.

The following table shows the results of these calculations. The first column gives the average density of the sun when its surface was at the orbits of the several planets. The second column gives the densities of the planets assumed to be at first fifteen times less dense than the sun. The third column shows the diameters of the nebulous planets. Hydrogen is used as the term of comparison for densities : thus the sun at the orbit of Mercury was thirty times less dense than hydrogen, and Mercury was four hundred and fifty times less dense than that lightest of gases. The diameters of the planets are in miles.

	DENSITY OF SUN.	DENSITY OF PLANETS.	DIAMETERS OF PLANETS.
	How many time less than hydrogen	How many times less dense than hydrogen.	In miles.
Mercury......	30	450	974,400
Venus	195	2,936	4,428,000
Earth	517	7,760	6,383,000
Mars.........	1,827	27,410	4,954,000
Jupiter.......	72,820	1,092,000	231,200,000
Saturn	448,800	6,732,000	284,600,000
Uranus.......	3,650,000	54,740,000	296,400,000
Neptune......	14,000,000	210,600,000	504,800,000

In looking over these columns of the densities of the sun and of the planets, we are astonished at these ancient conditions of matter, and we learn some valuable lessons. We are here presented with a connecting link between the two wide extremes of matter; matter in one condition as solids, liquids, and gases, and in another condition as those ethers whose vibrations produce light, heat, magnetism, gravity, and chemical action. The vibrations of light are 458,000,000,000,000 in a second, and they travel at the rate of nearly 200,000 miles in a second. What an inconceivably rare substance must it be! But is it more rare than the sun when expanded to the orbit of Neptune? According to the nebular theory, the sun was originally expanded a few thousand times farther than the orbit of that planet; but when enlarged only one thousand times farther; it was forty-five thousand billion (45,000,000,000,000) times less dense than hydrogen. These numbers may appear extravagant, but they are no more extravagant than many other numbers of science; for instance, those just mentioned about light, and also those of the distances of the fixed stars. While light travels at the rate of nearly 12,000,000 miles in a minute, it yet requires many thousands of years to come from the more distant stars visible by the telescope. If the nebular theory be true, then it seems that, as far as our conceptions go, the matter now composing the solar system was once as rare as those ethers forming light, heat, magnetism, gravity, and chemical action. Therefore as a solid gift by the nebular theory to our knowledge of " the physical constitution of matter," we have this. Just as in geology we trace the forms of plants and of animals, back little by little to a former very simple condition, so we trace back the matter of our globe little by little to a very ancient and simple state; not only inconceivably rare, but, as has already been shown, without those present modifications which we call the simple chemical elements. Just as geology presents us with connecting links between very different animal and vegetable forms now living, so the nebular theory presents us with a form of matter which, in point of extreme rarity at least, is as nearly as we can conceive a connecting link between hydrogen and the ether whose pulsations form light. Here is an unexpected illustration of the saying, *Natura non facit saltum.*

In looking over the column in the table, expressing the original diameters of the nebulous planets, we see a remarkable difference between the first four and the last four. This, as we shall soon see, unlocks the mystery why the four exterior planets have all the satellites but one, and why of the four interior planets, the earth alone has a satellite: Mars, Venus, and Mercury, were so small, that gravity had not the power to produce rings and satellites. On the surfaces of the four larger of these great globes, gravity was so very feeble, that we may neglect its precise amounts at those points in calculating their velocities of rotation, and consequently the velocities of their satellites. The inclined planes down which the matter of their equatorial zones descended, may be regarded as coming from infinite space. This is not the case with the four smaller nebulous planets. Gravity was considerable on their surfaces, and in calculating their velocities of rotation, and consequently their capabilities of having rings and satellites, their inclined planes must be taken as coming only from their original nebulous surfaces, providing we locate those surfaces by the densities in the table; that is, fifteen times less dense than the sun when expanded to their orbits. If their inclined planes were calculated as extending out to infinite space; a much greater velocity would be assigned to their rotations, and consequently to our moon, than is their due.

Having now approximately determined the heights of the inclined planes of the planets in the beginning, and knowing their masses, or the motive powers on those planes, we are prepared to calculate their velocities of rotation. In so doing, we find that some of them must have had velocities great enough to produce rings, and that those rings and their resulting satellites must have had the same velocities as the present velocities of the several satellites, allowing only a moderate loss of velocity by friction. That friction, as already shown, was an absolutely necessary element in the formation of all great globes of space. The following table exhibits the actual and the calculated velocities of all the satellites. The calculated velocities of all except our moon, are found in the same way as those of the planets. That of our moon is found by the formula $N-M$, also already employed for the planets, and explained in Appendix I., in which M represents the velocity of a fall from infinite space to the earth's original nebulous surface, and N the velocity from infinite space to the

VELOCITIES OF THE SATELLITES.

moon's orbit. Only two satellites of Uranus are given, because the others are so imperfectly known. The first is the column of actual velocities, the second of the calculated velocities, and the third of the retardation by friction. The velocities are in miles and decimals of a mile per second, except our moon's, which, on account of its slowness, is in miles per minute. The fourth column, for Saturn and Jupiter, shows the percentages of loss in velocity by retardation. The fifth column, for those two planets, shows how many miles they contracted in radius from time to time for a loss in velocity of one per cent. This fifth column was obtained in the same way as the fourth column in the corresponding table respecting the retardations of the planets.

	Actual velocity.	Calculated velocity.	Retardation.	Percentage of loss in velocity.	Miles of contraction in radius for loss of 1 per cent. in velocity.
Satellite of Neptune.	2.806	3.993	1.187	....	.....
Satellites of Uranus.					
4th	2.121	2.631	.51	....	.....
2d	2.452	2.532	.08	....	.....
Satellites of Saturn.					
8th	2.213	2.824	.611	21.7	6,455,500
7th	3.393	3.671	.278	7.6	179,560
6th	3.787	3.926	.139	3.6	60,990
5th	5.767	6.304	.537	8.6	54,880
4th	6.810	7.119	.387	4.4	23,120
3d	7.701	7.955	.254	3.1	18,151
2d	8.581	8.804	.223	2.6	14,815
1st	9.717	10.031	.324	3.2	11,156
Satellites of Jupiter.					
4th	4.928	7.484	2.456	34	3,660,230
3d	6.808	7.185	.377	5.3	89,056
2d	8.532	9.367	.835	9	28,166
1st	10.74	11.692	.952	8.1	19,345
Satellite of Earth.	37.93	39.11	1.18	....	.....

VELOCITIES OF THE SATELLITES. 249

The columns for the satellites of Saturn and Jupiter, which show the loss of velocity as due to gravity of one per cent., during certain distances of contraction in radius, follow the law of retardation formerly described in the nebulous sun. In approaching the denser regions toward the centre, the retardation becomes greater and greater. The retardations of 21.7 per cent. and 34 per cent. for the two exterior satellites, seem very large, as in the instance of the exterior planet Neptune. But they are in reality the smallest of all. They were the entire amounts accumulated from the beginnings of contraction when the nebulous planets were very large, as given in a former table. They amount on an average to only one per cent. for every distance of 3,600,000 miles in radius in one case, and 6,400,000 in the other.

There is an apparent contrast, however, between the amounts of retardation suffered by the satellites on one hand, and the planets on the other. The satellites lose one per cent. of their velocity, due to gravity, in a much smaller number of miles. Hence their retardation seems greater. But in reality it is much less. A mile, as a unit of measure, cannot be applied alike in both cases. A mile for the feeble gravity of the planets is a much greater quantity than a mile for the powerful gravity of the sun. Hence, to form a comparison, we must apportion the measure in distance to the masses of the bodies compared.

We may here pause for a moment and contemplate again some more inevitable consequences of gravity, acting on a contracting nebulous mass. Here are eight nebulous planets, and all are rotating as gravity should make them rotate. The velocities of their rotations, when a slight abatement is made for necessary friction, coincide at their different sizes with the velocities of their satellites! Therefore, at the orbits of these satellites, the centrifugal and centripetal

11*

forces were equal, and equatorial rings must have been separated from these nebulous planets to form satellites. The origination of these satellites, therefore, are so many more inevitable consequences of the action of gravity. This fact stands out in more conspicuous prominence when we see that the smaller planets have no satellites, and that our earth has only one. Mars, Venus, and Mercury, by reason of their diminutive sizes and masses, had not inclined planes high enough, nor a force of gravity working on these planes strong enough, after friction was overcome, to acquire a velocity that would separate equatorial rings. The earth is the largest of the four interior planets, and barely reached the volume and the mass to have a satellite, and hence it has only one. Here are several striking and unexpected coincidences between the action of gravity and very different facts. It is clear that by the action of gravity all the satellites have been separated from their parent masses the planets, the same as the planets themselves were separated from the sun.

What a wonderful piece of mechanism do we here behold! Counting planets, asteroids, and satellites, there are in the system more than a hundred round bodies. They all have two motions, and some of them three. They move with inconceivable velocities. They revolve around one another in a most intricate and in an eternal dance. And yet the whole can be accounted for by a single principle—the action of gravity on a contracting nebulous mass! The origination of the solar system is like a lock of the most intricate mechanism, and gravity is the single and only key to fit that lock! A key which fits and works a lock so complicated, must be the true one.

One obvious peculiarity of this mechanism is, that all the rotations and revolutions are in the same plane, and in the same direction—from west to east. This, as well as the peculiar velocities of the several orbs, and their necessary derivation or separation from one another, are so clearly coincident with the action of gravity, as to be truly admirable. The minor deviations from the same plane, and the inclinations of the axes of rotation to the orbital planes, and the anomalies of Uranus, can all be easily accounted for.

The reason why Neptune appears to us as yet to have but one satellite is, doubtless, on account of his vast distance. His distance is more than three times that of Saturn, whose last satellite was only recently discovered, and nearly twice that of Uranus, all of whose satellites have been seen by very few observers, perhaps by only one.

Hitherto we have attended only to the grand outlines of the solar system, and we have traced them as inevitable consequences of gravity acting on a given body. We will next examine the minor details of that system.

SECTION XXII.

WHY THE EQUATORIAL VELOCITY OF THE SUN IS LESS THAN THE ORBITAL VELOCITIES OF THE PLANETS; AND WHY THE EQUATORIAL VELOCITIES OF THE PLANETS ARE LESS THAN ORBITAL VELOCITIES OF THE SATELLITES.

AMONG the points of resemblance between the general solar and the special planetary systems, there is one very curious, in which they all seem to oppose the nebular theory. After a rotating nebulous globe has given off its last ring, and while it keeps on contracting, its velocity of rotation, urged on by the power of gravity, should increase; and therefore its equatorial velocity should be greater than the orbital velocity of the last ring and its resulting globe. But this is not the case with the sun in its general, nor with the planets in their special systems, as the following table shows:

	Equatorial velocity in miles per hour.	Orbital velocity of nearest revolving attendant.	
Sun...	4,564	110,725	Mercury.
Jupiter..	28,128	38,772	Nearest satellite.
Saturn...	22,441	34,986	" "
Earth...	1,040	2,276	The Moon.

Uranus and Neptune are not introduced in this table, because their rotations are not known. The cases above are very powerful in exhibiting an apparent contradiction to the nebular theory. And when, instead of making a comparison with their nearest attendants, we compare their present actual velocities with their calculated velocities, by the inclined plane theorem, this seeming contradiction stands out in stronger relief. This latter comparison is as follows. Mercury is not in the table, because of the uncertainty of its mass:

	Actual equatorial velocities in miles per hour.	Calculated velocities in miles per hour.	Number of times retarded.
Sun...	4,564	1,394,300	305
Venus...	1,050	23,680	22.55
Earth...	1,040	24,990	24
Mars...	523	12,610	24
Jupiter..	28,128	137,300	4.88
Saturn...	22,441	81,470	3.63
Uranus...	11,410?	45,040	4. ?
Neptune..		49,900	

The third column shows the proportion between the actual and the calculated velocities: thus the equatorial velocity of the sun is 305 times less rapid than it should be by the force of gravity. The earth's equatorial velocity is twenty-four times less rapid than is demanded by its inclined plane, and its power on that plane; and similarly, according to the third column, stand the cases of the other planets. The explanation of this apparent contradiction to the nebular theory greatly strengthens that theory. We have already seen how all the planets and all the satellites have been retarded in their orbital motions, and we have seen how that retardation was caused by friction. Retardation from friction has again been the cause of the slow equatorial motions of the sun and the planets. But why in these equatorial motions has the friction and the consequent retardation been so great? In the orbital motions the retardations were very moderate; here in the equatorial motions they are enormous. After the parting of the last rings, the equatorial velocities of all the nebulous globes, instead of increasing, wonderfully decreased! The reason is this: the retardation, as explained in the nineteenth section, was caused by the unrotating interior of the nebulous body. The surface rotated from causes already explained, while the particles of the great interior were contracting in radial lines toward the centre. There was nothing to cause them to rotate until they were acted on by the rotating exteriors. In the case of the sun especially, the *chief mass* of that body did not rotate until it had contracted within the orbit of Mercury. Hence so very small a proportion separated as rings, and hence also when a greater density was acquired, and when the moving exterior and the unmoving interior came in closer contact, friction grew to be far more powerful. Then with that more powerful friction there was no more acceleration of equatorial motion; on the contrary, that equatorial motion began to slacken, then it grew slower and slower, until it became as we now behold. The equatorial region of the sun, from moving 110,725 miles an hour, as at the orbit of Mercury, has had its speed so much retarded that now it moves only 4,564 miles an hour!

It will aid us to estimate the power of this friction by calculating the exceeding small inclinations of the planes down which the matter of the equatorial regions flowed.

If our sun now contracts in diameter the one-twentieth of a mile in a year, and this would be a rapid contraction, in my opinion, then the length of its inclined plane must be to its height as 1,600,000,000 is to 1. On such an extremely slight inclination friction would manifest its power nearly the same as on a level surface.

From the great retardation of the surface of the sun we arrive at the corollary that, when it extended to the orbit of Mercury, the rotating layer must have been comparatively thin. Had it been deep and heavy, and its momentum consequently great, it could not have been so much retarded.

These two principles, that of Retardation and that of Motion down an inclined plane, are the great keys to unlock the mysteries of the solar system. To show the influence of retardation on the present economy of our lives, we may pause to see how it has rounded the form of our globe, lengthened our days, especially our winter days, and ameliorated our climate. In the following table the first column shows the actual number of hours in a single rotation of some of the heavenly bodies; and the second shows the number of hours there might have been if the retardation had been a certain amount less—so much less, that the centripetal and centrifugal forces at their equatorial surfaces would have been nearly equal:

	Actual time of one rotation.	Possible time with less retardation.
Sun	670 hours 4 minutes.	3 hours 00 minutes.
Venus	23 " 21 "	1 " 27 "
Earth	23 " 56 "	1 " 20 "
Mars	24 " 37 "	1 " 26 "
Jupiter	9 " 55 "	2 " 52 "
Saturn	10 " 29 "	4 " 5 "
Uranus	9 " 30 "	3 " 24 "
Neptune		3 " 20 "

With these rapid rotations in the second column, the oblateness of all these great globes would have been such that the equatorial would have been double the polar diameters, and the icy climate of our present polar regions, with several months of midnight darkness every year, would have extended far down to low latitudes. The sun, so greatly flattened, would have given us far less light and heat, and his rapid motion would have precluded any examination of his surface or constitution. At the seasons of the equinoxes our days would have been only forty minutes long, and at the winter solstice here, in the latitude of forty degrees, I suppose there would have been but a very few moments of sunshine each day, perhaps none at all, as I have not the leisure to make the calculation.

SECTION XXIII.

WHY THE SUN HAS BEEN RETARDED IN ITS ROTATION SO MUCH MORE THAN THE PLANETS.

	Equatorial velocity in miles per hour.	Orbital velocity of nearest attendant.
Sun	4,564	110,725
Jupiter	28,128	38,772
Saturn	22,441	34,986

By this table we see that the difference between the equatorial velocity of the sun and the orbital velocity of its nearest attendant is very great, being as one to about twenty-seven. But the difference between the equatorial velocities of Jupiter or of Saturn, and the orbital velocities of their nearest attendants, is not so great, being less than one to two. Why this contrast between the sun and the planets? Evidently it is because the sun in its rotation has been so much more retarded. In like manner we perceive by the second table of the last section that the difference between the sun's actual rotation and his rotation calculated by the inclined plane theorem, is as 1 to 305; but the difference in the case of Jupiter is as 1 to 4.88, and in the case of Saturn as 1 to 3.63. Here we see again how much more the sun has been retarded than the planets. The nebular theory gives a satisfactory explanation of the difference in their retardations. The sun was formed of a comparatively well-collected nebulous mass. If orignally it was not round, it was yet massive in shape, and not scattered and straggling. Therefore, in assuming a round form under the influence of gravitation, it was only necessary for the prominent angles to move laterally into the depressions, and thus by surface movements alone the stable spherical shape was attained. The entire interior remained unmoved in this operation, and did not rotate. Hence the thin stratum of the rotating exterior was so greatly retarded by the deep unrotating interior. But the nebulous masses forming the planets had originally a very different form. They were long cylinders or prisms coiled around in the form of rings. Hence to collect under the influence of gravity into spherical forms, the movements or currents must take place not only on the surfaces, but more or less through the interiors. Hence a considerable portion of the interiors began to rotate by this primitive motion, and not through mere friction with the rotating exteriors. Hence the exteriors of the planets were not so much retarded by this friction on unrotating interiors as was the case in the sun.

SECTION XXIV.

THE DISTANCES OF THE PLANETS FROM THE SUN, AND OF THE SATELLITES FROM THE PLANETS.

IN surveying the distances of the planets from the sun, and of the satellites from the planets, we perceive another strong point of resemblance between the general solar and the special planetary systems. In approaching toward the sun, the planets are arranged nearer and nearer together as far as Venus, and the distance to the next planet, Mercury, is wider. The same is true of the satellites: the distances from one to another become less and less until toward the last; and when very near the planets, these distances are lessened very little or not at all. In the cases of Uranus and of Saturn the last intersatellite space is wider than the one immediately preceding. While thus looking broadly at the facts of the distribution of both planets and satellites, we see that the same influence has been at work, crowding them closer and closer together, until at length that influence has been counteracted; there is no more approximation, and soon no more planets or satellites have been formed. The approach to regularity in this planetary arrangement is so remarkable as to have led to the discovery of the first four asteroids; and many attempts have been made to give expression to the planetary distribution by mathematical law. This evident approximation to law is called Bode's law. Without entering into these attempts to find a strict mathematical law, our object will be to discover what influence in the nebular condensations and in the production of rings has caused this peculiar arrangement.

In the following tables are given the radii of the nebulous globes having satellites, and the distances of the farthest satellites from the planets. The third column of figures shows how many times these radii are greater than the distances of the farthest satellites. These radii are calculated on the supposition that the nebulous planets, in their spherical forms in the beginning, had a density fifteen times less than the sun when enlarged to their orbits. The rings must have been much less dense, and they must have contracted greatly before attaining to globular shape.

	Radii in miles.	Distance of farthest Satellite.	How many times greater.
Neptune......	252,400,000	225,000	
Uranus.......	148,200,000	1,570,000	94.4
Saturn.......	142,300,000	2,414,000	58.9
Jupiter.......	115,600,000	1,152,000	100.
Earth	3,196,500	237,000	13.5

 * Neptune's farthest satellite, I suppose, has not yet been discovered, and therefore his place in the third column of figures is not computed. By this table it appears that our earth, before giving off a ring, had contracted from a distance 13.5 farther than the moon. Jupiter had contracted one hundred times, Saturn fifty-nine times, and Uranus ninety-five times the distance of their farthest satellites, before producing a ring. And then when rings began to be formed, they were in all cases formed very slowly at first, and increased in frequency or nearness together until the last or next to the last. The reason why rings were produced so very slowly at first, is the same why previously through so vast a distance of contraction they were not produced at all. Why were not rings produced in the very commencement of rotation? Why wait so long? Why first contract twelve parts out of thirteen, or ninety-nine parts out of one hundred, before a single ring could be formed? After the first ring, why so much sooner a second ring, and still sooner a third, and then a fourth after a shorter space than any of the preceding, and so on until the last or next to the last satellite or planet? Nothing can be more simple than the explanation of this curious phenomenon, if we apply the facts about retardation given in a former section. In the beginning of the rotation of a nebulous globe, the rotation is only of a thin layer on the surface. It is not enough to produce oblateness, the extra expansion on the equatorial zone, which is the budding of a ring. The rotation is constantly retarded by friction on the undisturbed interior.

That interior even before it rotates must be compressed in an oblate form, and this act of compression impedes the velocity of the exterior. From this velocity the centrifugal force may nearly equal the centripetal force; but unless the two be exactly equal, no rings can result. Hence it is that twelve-thirteenths or ninety-nine hundredths of the life of a nebulous globe are passed away before a ring can be separated. If the nebulous globe be not very large, no ring can be parted, as in the case of the three smallest planetary globes, Mars, Venus, and Mercury. But after one ring has been formed, then the globe is at an age to produce another in a short time, and a third and a fourth still sooner. This is because the rotating layer has been deepened and hastened in its velocity up to the point where the centrifugal and centripetal forces are equal. Complete oblateness has been assumed, making the equatorial about double the polar diameter. The rings are therefore nearer and nearer as the obstructions to the rotating interior are more and more out of the way. Even if the same amount of obstructions to the rotating exterior existed, its increasing thickness and velocity would make these obstructions more and more easily overcome. This increasing velocity is of prime importance, and alone sufficient to determine the greater frequency of the rings. I refer not simply to the proportionally greater increase of *vis viva* of a greater velocity for overcoming obstructions, but chiefly to the fact that the velocity acquired down an inclined plane produces a centrifugal force which *increases more rapidly* toward the centre of the sun than the centripetal force. Hence, in this gain of the centrifugal over the centripetal force, we see the gain in power to produce rings, and consequently a gain in the frequency of ring births.

But at length a new epoch arrives and a new phase of action. In the course of contraction a density is acquired, when, by friction, the whole mass of the unrotating interior begins to feel the force of the rotating exterior. In the case of the sun that interior had been condensing and settling in radial lines. From long anterior to the formation of Neptune, it was the chief part of the mass of the sun which had been settling in radial lines. While the light surface rotation had been going on, the dense interior had not rotated. Now a point of density is reached where friction grates strongly on the whole of that heavy inert interior, and as a consequence the last ring is a little delayed beyond its time. Therefore, Mercury is some two millions of miles farther from Venus than Venus is from the earth, and therefore also the last satellites of Uranus and Saturn were given off after a wider space. At last the centrifugal force, in consequence of the friction on the entire interior, becomes a little weaker than the centripetal force, and then not another planet is born. Therefore, the fact of retardation, an unavoidable occurrence in most nebulous globes, tells why the production of planets and satellites is delayed so long, and in some cases prevented altogether. It tells why they are formed slowly at first, then more rapidly, then a little more slowly again, and then an entire cessation of planetary births. Dynamical laws, gravity, and gravity conserved as inertia, the laws of contraction and friction, are the same in all nebulous globes, whether they be suns or planets, and hence we behold the same general law of distribution among both planets and satellites.

There is another and a numerical method of illustrating the cause of the peculiar distribution of the planets and satellites. In the table of the nineteenth section (p. 234) it appears that there is a loss of velocity due to gravity not far from the same amount in all the planets; ranging from eleven per cent. in Saturn to about six per cent. in Venus. This average of about eight per cent. of the force of gravity was expended, not in giving velocity to the planets, but in friction on the unrotating or slowly rotating interior. Hence, we see that

WHY THE SATELLITES TO EXTERIOR PLANETS. 261

when about eight per cent. of the force had been expended in friction on the slowly rotating interior, then the exterior on the equatorial zone had gained velocity enough to be separated as a ring. Now a loss of velocity due to gravity, say of eight per cent., must occur in a much shorter space as gravity increases in power. Gravity increases in a duplicate proportion; therefore the interplanetary spaces decrease in a simple proportion—or nearly in a simple proportion—as two is to one. These percentage numbers representing the loss of velocity or force, varying from eleven to six, would, I suppose, have been in all cases the same, and the interplanetary spaces would have followed a regular law, had it not been for inequalities of size and density in the strata of the nebulous sun. These inequalities also prevented a regular law in the masses of the planets. The more rare the stratum, the more oblate it would be, and its separation would occur proportionally sooner than that of the denser one next below.

SECTION XXV.

WHY THE FOUR EXTERIOR PLANETS HAVE ALL THE SATELLITES BUT ONE; AND WHY OF THE FOUR INTERIOR PLANETS THE EARTH ALONE HAS A SATELLITE.

To solve these questions, we must take into view the sizes of the planets when in their original nebulous conditions. Their diameters in miles were as follows, if calculated at a density fifteen times less than that of the sun when enlarged to their orbits:

Diameters in miles.	Diameters in miles.
Mercury........ 974,400	Jupiter....... 231,200,000
Venus.......... 4,428,000	Saturn....... 284,600,000
Earth.......... 6,393,000	Uranus....... 296,400,000
Mars........... 4,954,000	Neptune...... 504,800,000

The contrast between these two sets of planets, in point

of size, while in their primary nebulous conditions, is truly wonderful. This contrast is equally strong when we compare their original densities, which were as follows:

How many times more rare than hydrogen.		How many times more rare than hydrogen.	
Mercury	450	Jupiter	1,092,000
Venus	2,936	Saturn	6,732,000
Earth	7,760	Uranus	54,740,000
Mars	27,410	Neptune	210,600,000

A still more pertinent point of contrast is the degree they have been retarded in the velocity of their rotations. When we compare the present velocities of their equators with the velocities they ought to have by the inclined plane theorem, we find that Mars has been retarded twenty-four times, and Saturn only three and a half times; that is, they would now have had an equatorial velocity so many times greater, but for retardation. The tabular statement is as follows:

Number of times retarded.		Number of times retarded.	
Venus	22	Jupiter	4.88
Earth	24	Saturn	3.63
Mars	24	Uranus	4. ?

The cases of Mercury and Neptune cannot be ascertained for want of data. The reason why the interior planets were retarded in their rotations so much more than the exterior, was, I suppose, on account of their greater densities and smaller sizes. But this need not be discussed here, as the facts of their retardation are so plain. On account of this principle of retardation, the exterior planets contracted their radii from fifty-nine to one hundred times the distance of their farthest satellites before producing a ring. Then, after producing rings for a certain period, their rotating exteriors were retarded anew from closer contact with the dense unrotating interiors. But before the smaller interior planets could, in the first place, contract so great a distance in order to produce rings, their rotating exteriors came directly in contact with their dense unrotating interiors, and hence they could part no rings at all. Therefore, on account of the primary retardation which the external layer had to encounter from the beginning, and the final retardation by the interior core, and also from the much shorter distance for contraction, the interior planets could not gain on their equatorial surfaces a velocity making their centrifugal equal to their centripetal forces.

But why of the four interior planets has the earth alone a satellite? The answer is plain, by viewing the table of their nebulous diameters. There we see that, in the beginning, while in their nebulous condition, the earth was larger by far than the other three interior planets. It was so much larger, and its mass so much greater,—in other words, its inclined plane was so much higher, and the power on that plane so much stronger, that a velocity could be acquired making the centrifugal equal to the centripetal force, after overcoming all retardation by friction. This could barely be done, and hence it has but one satellite.

SECTION XXVI.

WHY THE SUN'S NEAREST ATTENDANT IS SO FAR OFF—THIRTY-FIVE MILLIONS OF MILES; AND WHY SATURN'S NEAREST ATTENDANT, HIS INNER RING, IS SO NEAR, ONLY ABOUT TWELVE THOUSAND MILES FROM HIS SURFACE; AND WHY THERE IS NO PLANET BETWEEN MERCURY AND THE SUN.

THE reason of the great difference between the sun and Saturn in the distances of their nearest attendants, is due to the difference between their retardations. The connection between the distances of the nearest attendants and the amounts of retardation will be seen by the following table:

How many times retarded.	Distance in miles of nearest attendant.
Sun 305	35,000,000
Jupiter 4.88	269,800
Saturn............ 3.63	50,000 Inner ring from centre.

By retardation the sun's equatorial velocity is three hundred and five times less than it would have been, and Saturn's only three and a half times less. Here we see plainly the reason why the formation of rings was arrested so far from the sun's surface. When the ring resulting in Mercury was parted, the sun's equatorial velocity was 110,000 miles per hour. By retardation it was reduced to 4,564 miles per hour. This was an enormous reduction. How, then, was it possible for another ring to be produced inside of Mercury? For the production of another ring, a much greater velocity than that of Mercury was necessary; and that greater velocity not being allowed by the principle of retardation, another planet interior to Mercury was impossible. If from Mercury's orbit we measure a distance toward the sun equal to the space between Venus and Mercury, which is 33,000,000 miles, we arrive at or very near the sun's surface. Long before the sun's surface by contraction reached that point, his equatorial velocity must have been very slow—by far too slow for parting another ring.

Saturn's retardation was the least of all, and hence his last ring was nearest of all. Jupiter occupies an intermediate place in retardation, and the distance of his last ring was accordingly intermediate.

SECTION XXVII.

THE PECULIARITIES OF THE SATURNIAN SYSTEM.

THE system of Saturn, while preserving the general characteristics of the general solar and of the special plane-

THE SATURNIAN SYSTEM. 265

tary systems, differs very strongly in some six or eight particulars. They are as follows:

1. An attendant so near his surface. His nearest ring, the last discovered and semi-transparent, is only about 10,000 miles from his surface, and therefore only about 47,000 or 50,000 miles from his centre. The distances between the centres of the other planets and their nearest attendants are as follows: Earth, 237,000; Jupiter, 269,800; Uranus, 226,000; Neptune, 225,000. These four distances are remarkably near each other, while that of Saturn from his attendant is about one-fifth of the average of these.

2. An attendant so very far off. His farthest satellite is double the distance between Jupiter and his farthest satellite.

3. The large number of his attendants, three rings and eight moons—eleven in all. This is nearly double the number attending Uranus, and three times the number attending Jupiter.

4. The nearness together of his attendants.

5. His great oblateness.

6. Three of his rings remain unbroken.

There are two other peculiarities which may be called theoretical, as follows:

7. The nearness of his equatorial velocity to the theoretical velocity by the force of gravity.

8. When compared with the other exterior planets, his distance of contraction before producing a ring was much less. Before producing a ring, Jupiter contracted one hundred times the distance of his farthest satellite; Uranus, ninety-five times; Saturn, only fifty-nine times. Let us now examine whether the principles we have already announced can give a satisfactory explanation of the origin of these peculiarities.

Retardation I have already shown to be the cause why a nebulous globe contracts so far without producing a ring;

in one case ninety-nine hundredths of its whole distance. From the same cause, also, the sun ceased to produce rings during the last 35,000,000 miles of its contraction. Now it is evident from comparing Saturn's actual equatorial velocity with the orbital velocity of his nearest satellite, and also with his calculated velocity as due to gravity, that this planet has suffered less retardation in his rotation than the sun or the other planets. Therefore, *à priori*, we should expect that this planet would have rings farther off than any other planet, and would have them nearer together, and consequently a larger number. From the same *à priori* views, we would necessarily look for greater oblateness, because if its velocity of rotation were more retarded, its oblateness must be diminished. Thus it appears that the first five of these peculiarities are traceable to the same cause, retardation. The nearness of the rings of Saturn to his own body, nearer than the rings of any other planet, is the reason why they remain unbroken. The nearer they are, the more strongly they are embraced by the influence of two opposing forces, the centripetal and the centrifugal. They are held as within the jaws of a mighty vice, and the feeble force of gravity of his moons has not been able to cause their rupture. The gravity of his moons is rendered especially feeble by their want of density. One of Jupiter's satellites is fifty times less dense than our earth, and as Saturn has only half the density of Jupiter, their satellites probably compare in a similar proportion. The two remaining peculiarities of Saturn, the seventh and eighth, not being visible, may be called theoretical. They are so obviously dependent on the small amount of retardation, as to require no additional remarks.

SECTION XXVIII.

WHY THE PLANETS ARE SO MUCH MORE DISTANT FROM ONE ANOTHER THAN THE SATELLITES.

THE distances between the planets are measured by millions and hundreds of millions of miles; those between the satellites by only a few thousands of miles. It first strikes us that there is a cause for this arrangement, in the fact that the solar system is so much larger than the planetary systems. But this is a mere analogical dream, pointing out no connection between cause and effect, and indeed pointing out no cause at all. If the planets had been so large that their rings must necessarily have occupied the whole of the interplanetary spaces, then there would appear a real cause. But, from our calculations of the dimensions of those rings, it is clear that they did not occupy the whole of these spaces.

The principles already laid down and applied to other questions, admit an easy application to this. Why did Saturn have his rings nearer together than any other planet? Less retardation. For the very same reason all the planets must have rings nearer together than the sun. Their retardations according to calculation were much less. The sun's equatorial velocity is now three hundred and five times less than it ought to be by the calculations as due to gravity. That of Jupiter is only 4.88 times, and that of Saturn only 3.63 times less. We have seen that for 35,000,000 miles between Mercury and the sun, there is no planet because of retardation; that if there had been no retardation, the planets would have been close together from the beginning of nebular contraction to its end; and if there had been a sufficient amount of retardation, there would have been no rings or planets at all. Therefore we may conclude that this obviously great amount of retardation in the sun was the cause why the planets are so much farther asunder than the satellites.

SECTION XXIX.

THE PECULIARITIES OF THE MUNDANE SYSTEM.

The planetary system of the earth has the following peculiarities:

1. The earth has only a single satellite, while the other planets with satellites have several. Neptune is not here taken into account, for all its satellites, as I suppose, have not yet been discovered. The nebular theory renders the existence of other satellites to that planet in the highest degree probable.

2. The moon moves more slowly by far than any other celestial body, only 2,276 miles per hour. The slowest satellite of Saturn moves about 8,000 miles an hour; the slowest of Jupiter about 18,000; and Neptune, the slowest planet, has a velocity of 12,570 miles per hour.

3. In proportion to the mass of the earth, the moon is by far more distant from its primary than any other satellite. This is shown by its slow motion.

4. The mass of the moon approaches more nearly to the mass of its primary than any other satellite. It is about the $\frac{1}{84}$ part of the earth. Jupiter's largest satellite is only the $\frac{1}{11300}$ of the mass of Jupiter, and his smallest is the $\frac{1}{518000}$. Jupiter's own mass, larger than that of all the other planets and satellites together, is only the $\frac{1}{1048}$ of the sun's mass.

5. The slowness of the moon's rotation, only once in a month, and so as to keep the same side always turned toward the earth. From theoretical considerations, and

from observations not rendered as clear as can be desired, this is the case also with the satellites of Jupiter and Saturn.

The first of these peculiarities has already been accounted for in the twenty-fifth section. The second, third, and fourth peculiarities may all be shown to have originated from one and the same cause. We have seen that the sun was retarded in its rotation so much more than the planets, because originally it was, as far as shape is concerned, a more compact mass; whereas the planets were collected from long cylinders or prisms coiled round in the shape of a ring. In this process of collection, their interiors received a motion from the very first, and hence did not so much retard their exteriors. The earth must be supposed to have been composed chiefly of two layers: an interior portion or layer not much different from that of the other smaller planets, and a deep exterior layer very little retarded, and consequently rotating freely. From this unusually free and unretarded rotation of the deep exterior layer, would result the fact that a ring would separate sooner and of greater size than usual. Therefore it would be more *distant* from the centre of its primary, and therefore also proportionally *slower* than the other satellites. For the same reason also, namely, the unretarded rotation of the *deep* exterior layer, the ring might be proportionally larger than usual, and hence the large size comparatively of the moon. We need not now inquire why a nebulous globe should collect together in such a ring as to have its exterior layer deeper and retarded less than usual. The possibility of such a case is evident, and that is enough for our present purpose in accounting for these three peculiarities.

The fifth peculiarity, the very slow rotation of the moon, and its presentation of the same side always toward the earth, has been accounted for by Mayer, in his "Celestial

Dynamics," in this manner. He assumes that the moon when in a liquid condition had a more rapid rotation than now; that from the earth's gravity a larger tide-wave arose on the side toward the earth; that from the moon's rapid rotation, this tide-wave was left behind, a little like our own ocean tide-waves, and hence really existed in some degree on the side of the moon departing in its rotations from the earth; that therefore this tidal protuberance was like a mountain on that side, and being attracted by the earth, it continually opposed and retarded the rotation, until this tidal protuberance was brought permanently toward the earth; and then, of course, the period of its rotation first corresponded with its period of revolution. Soon after this was published in Germany, and before its appearance in our language, I came to the same conclusion independently. But my conception and description were in a way very different, and as I think superior to that of Mayer. Copied from my old manuscript, it is as follows:

"The *two* tide-waves of the moon were always the one on the side toward the earth, and the other precisely on the opposite hemisphere. As the moon moved round on its axis from west to east, the tide-wave moved in an opposite direction from east to west, as now on our globe. When rigidity began in the solidifying surface of the moon, this rigidity opposed the advance of the tide-wave, and held it back. Just in proportion to the power of this rigidity did it delay the flow of the tide-wave, and thus counteract the tidal influence coming from the earth. Here there were two opposing forces: the rotating force of the satellite acting through its rigidity against the force of gravity emanating from the superior planet. This antagonism was complete; and in such a contest there could be but one result. The force of gravity of the planet is indestructible as far as we can perceive, but the rotating momentum of the satellite was and still is a definite quantity, capable of being destroyed by an opposing force. That opposing force, a very great one, was every moment at work destroying little by little the rotating momentum, until at last it was so far destroyed as only to turn the moon round on its axis once in one revolution around its primary."

Mayer's conception of a single tide-wave on the departing side of the moon, and that operated upon by the earth's gravity, as upon a huge mountain, seems to me objectionable. The moon like the earth must have had two tide-waves, and as they were on precisely opposite sides, the gravity of the earth, as far as it affected the moon's rotation, must have been the same on both, or so nearly the same, that the difference on account of their different distances must have been inconsiderable. Therefore the rigidity in the revolving body opposing the tide-waves, seems to me to have been the real agency that retarded the rotation.

The rotation of the earth and all the planets must have been affected by the same tide-wave agency. Here, therefore, we come to a second cause for the great fact of retardation which we have seen to have had so powerful an influence in the formation of the solar system, and which must be again taken into account when we explain the regular forms of the nebulæ; meaning now by this term the clusters of stars which, from vast distances, have a nebulous or cloudy appearance. The inquiries immediately arise, Why have not the planets been retarded by these tide-waves so as to present the same side always toward the sun, or, in the case of the earth, toward the moon? And, what is the proportion in amount of power and practical influence between these two causes of retardation? To the first question the following facts must be adduced.

1. By reason of differences in distance, the tidal influence of the sun on the planets is far less than that of the

planets on the satellites. The mass of the earth being eighty-four times greater than that of the moon, its gravitating influence in producing a tide-wave in the moon is just so many times greater than that of the moon in producing a tide-wave in the earth. The small superficial gravity of the moon, and its shorter diameter, must be taken into account; but these two tend to counterbalance each other. But even the moon's tide-producing power on the earth is greater than that of the sun; how much greater therefore must be the tidal influence of the earth on the moon than that of the sun on the earth!

2. The rotating momenta to withstand the influence of the tide-waves were incomparably greater in the planets than in the satellites; greater on account of their masses, and greater on account of their rotating velocities.

From these two causes, their much smaller tide-waves, and their much greater momenta, the planets have been far less under tidal influence than the satellites. When in a nebulous condition with very great diameters, their tide-waves must have been exceedingly high; but from the want of rigidity in the nebulous globes, I do not suppose that these waves then had any appreciable influence on their rotations. Neither do I suppose that the rotation of the earth at the present moment is affected by any tidal influence on account of its fused interior beneath the solid crust. Therefore the period for the retardation of the earth and the other planets by tidal influence, must have been chiefly during the first solidification of the outer crust, while it was yet thin and yielding, and before it became unsusceptible of interior tides.

In answer to the other question on the proportion in amount of power and practical influence between the two causes of retardation, very little can be said, because numerical data, as far as I can perceive, are not attainable.

As the result of my own reflections, I can simply say that, in my estimate, the influence of the tide-waves, as a cause of retardation, has been very much inferior to that of the unrotating dense interiors of the celestial bodies; though in this respect there may not have been a uniformity among the various cases.

Mayer has been the first and only writer, as far as I am aware, to describe the contest now going on between the tide-waves of the ocean and the cooling of our earth. Subsequently, and before his publications came to my knowledge, I thought out the subject independently. In this case there is not so wide a difference between his reasoning and my own as there appeared in the former; still the juxtaposition of the two statements, I hope, will not be regarded as out of place. Copied from my old manuscript, my views were as follows:

" The ocean of waters on our earth is now subject to tidal waves, and these, running contrary to the rotation of the earth, and falling with all their great momenta against continents and islands, ought gradually to lessen the rate of the earth's rotation, and make the days longer. Mathematicians have expressed surprise that this effect has not been detected.

" The interior heat of our globe is slowly radiating away, not only from three hundred volcanoes and innumerable hot springs, but even through the rocky crust both of the ocean bed and of the uplifted land; and hence there must be a contraction of our planet, and consequently we should look for a more rapid rotation on its axis, and a shortening of the days. Geologists have been puzzled to find from observation no such effects.

" But it is plain that these two anticipated effects, that of the tidal wave and that of the escape of heat from the interior of our globe, must counteract and probably evenly balance each other, and preserve the length of our day without any appreciable change."

SECTION XXX.

WHY THE PLANETS AND SATELLITES ROTATE FROM WEST TO EAST.

IT is easy to understand why the planets revolve around the sun from west to east. But why they rotate on their axes in the same direction is not so clear; a closer attention is needed, which we will now bestow. It depends entirely on the fact that the exterior of a ring revolves with the same angular velocity around the sun as its interior. And this latter fact depends again on two causes: First. Even if the exterior of a ring had a tendency to revolve more slowly, in the angular sense, than the interior, still the exterior would be carried along by friction, side by side, equally with the interior, and with the surface of the nebulous sun, before separating. For instance, before the ring of Neptune was separated from the interior nebulous mass, its contact with that mass was through the entire round of the immense orbit of that planet, and friction from this extensive contact would tend to carry forward the whole ring, both exterior and interior, with the same angular velocity. Friction in that very rare substance was slight indeed, but it must have had a very small effect, and that was sufficient. We see the effect of friction in the fact that the calculated velocities of all the planets are a little greater than their actual velocities. Secondly. We have already seen that the exterior of the nebulous sun rotated with a far greater angular velocity than the interior; and that in fact, through the greatest extent of its contraction, the interior did not rotate at all. Hence, the exterior of the ring, on the last supposition, must have had an angular velocity equal to the interior. Even if it should tend to fall behind, friction would come in aid.

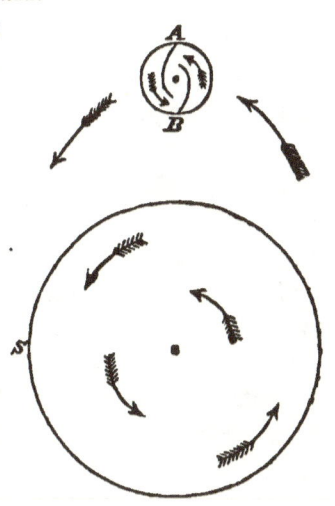

When a ring broke, from causes already explained, the opening became wider and wider. After a time the ring was merely a semi-circumference; then it decreased to a quadrant; then still shorter; and at length by the force of gravity it was fashioned into a globe. In revolving around the sun, S, the angular velocity of the exterior side of the globe at A, was equal to that of the interior side at B, and consequently its linear velocity was greater. Hence, in contracting toward the centre, the exterior and the interior sides would not move in straight lines toward each other. The former, A, moving more rapidly than the centre, would go before the centre; and the latter, B, moving less rapidly than the centre, would fall behind the centre. Both points, A and B, would constantly approach the centre, but in spiral lines, and once having begun this spiral motion, it must be continued; the rotation must keep on in the same direction, and be continually increased, as has already been shown. But this rotation, evidently, from an inspection of the figure, must be in the same direction as the revolutionary motion, from west to east.

The satellites are related in the same way to the planets as the planets are to the sun, and the rotation of all must be in the same direction. This is the case with the satellites of the Earth, of Jupiter, and of Saturn. The others are too far off for observation. The cause of the retrograde motion of Uranus will be explained in another section.

But it has been supposed that the exterior of a ring cannot have the same angular velocity as the interior, on account of the great breadth of the rings. This is on the principle that a more distant planet must have a less linear and angular velocity than one nearer the sun. This supposition would be true, providing the rings were broad enough. Therefore it is necessary to ascertain the original breadth of the rings; that is, their dimensions in the direction of the sun's radii at the equator. This we may call the radial dimension, and the one parallel to the axis of the sun's rotation we may call the polar dimension. But can we ascertain the sizes and shapes of those ancient bodies so long ago changed into spheres? By dynamical laws, and by mathematical calculation, we can reconstruct and restore those primitive palæological forms in astronomy, the same as in geology, from petrified fossil remains, we can reconstruct the primitive animals of our globe whose lives have long since been extinct. The process is this: We have already ascertained the diameters of the nebulous planets when of the same density as the sun enlarged to their orbits. From their sizes find their cubic contents. Divide their cubic contents by the lengths of their orbits around the sun, and this will give the sections of the rings. Divide the section of a ring by its polar dimension, and the quotient will be the radial dimension. The two difficulties in this process are to find the densities of the rings as compared with the average of the sun, and their polar dimensions. By sufficient time and labor, we can approximate quite nearly to both these elements. We may assume, as a first approximation, that the rings separated from a zone of the nebulous sun in breadth one degree on each side of the equator; then the polar dimension of the rings will be equal to two degrees of latitude; more correctly, equal to twice the sine of one degree. It may have been wider, but not very greatly, because the sun was extremely oblate, and came to a comparatively sharp edge. At higher latitudes also the centrifugal force decreases very rapidly—inversely as the squares of the distances from the axis of rotation. Therefore, assuming two degrees to be the polar dimension of the ring, and its density equal to the average of the sun, we obtain the radial dimension. Afterwards we can alter them by multiplication as we think best. For instance, if we come to the conclusion that the rings were ten times less dense than the average of the sun, then we may multiply the polar dimension by five, if that be our conclusion, and the radial dimension by two. The following table exhibits the two dimensions of the rings on the basis that their densities equalled that of the nebulous sun enlarged to their orbits, and that their polar dimensions were one degree on each side of the equator:

	Polar dimensions.	Radial dimensions.
Neptune	96,650,000	2,200,000
Uranus	61,600,000	1,488,000
Saturn	30,630,000	5,074,000
Jupiter	16,740,000	9,514,000
Mars	4,890,000	1,103
Earth	3,210,000	5,632
Venus	2,322,000	3,314
Mercury	1,242,000	2,232

The only rings having a considerable radial dimension are those of Saturn and Jupiter. If we regard them as being ten times less dense than the average of the sun, and their polar dimensions occupying five degrees on each side of the equator, then their radial dimensions would double those in the table. Even then, in view of friction, and of the fact that the exteriors of the rings had originally a greater angular velocity than the interior of the nebulous mass, I believe that the angular velocity of the exterior of the ring equalled that of the interior,—or so nearly equalled it as to have an equal or greater linear velocity.

In these calculations, the fact must be borne in mind that each particle of a gaseous ring tends to revolve in a great circle around the centre of the nebulous sun, and this must operate to reduce the polar dimension of the ring; to mould them ultimately into the shapes of those of Saturn, if they exist long enough without breaking. In this way we may account for the extreme thinness of Saturn's rings.

SECTION XXXI.

THE INCLINATION OF THE AXIS OF ROTATION TO THE ORBITAL PLANES; THE INCLINATION OF THE ORBITAL PLANES OF SECONDARIES TO THE EQUATORIAL PLANES OF THEIR PRIMARIES; AND THE RETROGRADE MOTION OF THE SATELLITES OF URANUS.

From the principles already announced, it would seem, at first view, that the orbit of every planet should lie in the plane of the sun's equator, that their axes should all be perpendicular to their orbits, and that the orbit of every satellite should lie in the plane of the equator of its primary. Thus among all the members of the solar system, their axes should be parallel, and the planes of all their orbits and all their equators should be one. We would naturally, without much reflection, form such an opinion of the working of the nebular theory, though it would be not only a very superficial, but a very wrong opinion. It would leave out of view some of the most important principles of the universe. It would be the same as saying that all acorns are alike, and if all be planted in the same field, with the same soil, the same sun shining upon them, and the same air fanning them, then every oak they produce must be alike. Instead of this, there would not only be no two trees, but no two leaves alike. The world is constructed upon principles that bring an infinite diversity out of unity. If we let fall a dozen pebbles on the floor ever so often, they would never lie twice in the same relative positions. If all matter were distributed through space, with an approach toward equality as nearly as any thing we know, still, in contracting and breaking up into separate masses, no two masses would be alike; and while their unity would be seen in taking the globular shape, and entering upon rotations, the systems of globes formed by each would be different. The acorns, in truth, and their positions in the ground, would not be alike. Neither could we, by the most refined machinery, make twelve pebbles alike, nor drop them on the floor twice in the same manner. And matter diffused through space, however equally, by the same mechanical principles, would still have inequalities; and chemical action, causing contraction, would have its variations as well as the sprouting of acorns. Conceive of things as we may, we can never get clear of the great fact of variety in unity. That is God's fiat: "the way of the Lord." The union of all the currents into one on the surface of a nebular globe could never obliterate all traces of the individual currents. The impulses of each one would still exist in their blendings, and there would always be surgings to and fro, and irregular boiling up in the whirlpools down the inclined plane. When a ring parted, it would carry with it the waves—the foaming billows on the surface of the equatorial zone; and as it broke and collected into a new rotating globe, the irregular momenta would go on. How could all these impulses flow together, and form an axis absolutely invariable? That would be something supernatural, a miracle. It would be like dropping a needle on a steel plate so perpendicularly that it would stand erect, evenly balanced on its point. Nothing can be immovable. All is in motion; and therefore the axis of every rotating globe is in motion. Perhaps, like pendulums, they incline now this way, and now that way, requiring hundreds of millions of years for a beat. The axis of our earth now points to a particular star. We can tell when it will point more than forty-six degrees another way. The axis of the sun can be no exception. The plane of his equator was once in the general plane of his planets, but it could remain neither there nor in any other immovable position. More than a hundred millions of years have elapsed since the sun parted the ring of our globe. We know this from the teachings of Geology on the antiquity of the earth. Since then his axis has altered about eight degrees; that is, one degree for every 12,000,000 years. Does this seem absurdly slow? No; we must learn to correct the ideas of our childhood. As Natural Philosophy teaches us the wonders of velocity, that of light being twelve million miles in a minute, so our theory points to a wonder equally great in the opposite direction, one angular degree in twelve million years! In this latter case, however, we cannot be definite; instead of twelve millions, it may be many more. We have as much, and even more, reason to wonder that the orbit of our earth is so little out of the plane of the sun's equator, than that it is so much out of that plane. Because the orbit of our earth is seven or eight degrees out of the plane of the sun's equator, that is no reason why the two planes did not coincide some hundreds of millions of years ago. Because the axis of our earth now inclines so that the plane of its equator is twenty-three degrees out of the plane of its orbit, that is no reason why the two did not coincide " in the beginning." When millions of years are required for the motion of only one degree, such motions are beyond our appreciation and beyond our calculations. But we cannot say a single word against the existence of such motions when by mechanical principles we see a probability for their cause, and when in the solar system we see facts attributable to their effects. If the axis of our earth has moved more than twenty-three degrees, then, for aught we can tell, it may yet move a little more than ninety degrees, and then it will be in a position similar to that of Uranus. But Uranus, by reason of extreme oblateness, and by his own motion while in a nebulous state imparted to his satellites, has carried his satellites along with him. Hence, we can account for their retrograde motion. But Uranus is nineteen times farther from the sun than our earth, and therefore is probably nineteen times older, and has had so much more time to change his axis.

It is wonderful how the facts respecting the positions both of the axes and of the planes, in the general solar and in the special planetary systems, support the view that the slow movements of the several axes prevent them from being parallel, and prevent the several planes, both equatorial and orbital, from coinciding as one. Let us look at these facts.

1. If the axis of our earth had at first been at right angles to the plane of its orbit, then at that period the plane of its equator would have coincided with the plane of its orbit; but with a very slight motion of the axis in the lapse of millions of years, there would have been an angle of a few degrees between the two planes. Then came the period for giving off the ring of the moon, and hence the plane of the orbit of the moon is now at an angle of five degrees from the plane of the earth's orbit. The position of the orbit of the moon therefore may mark the former position of the plane of the earth's equator. The inclina-

tion of the earth's axis has been going on, and the plane of its equator is now removed more than twenty-three degrees from that of its orbit. The movement of the plane of the moon's orbit has probably been arrested by perturbing influences.

2. In accordance with this view, we behold that the planets which are nearer the sun, and consequently the last produced, are also nearer the plane of his equator. The plane of Mercury's orbit almost coincides with the sun's equator; that of Venus is only three or four degrees removed; and that of the earth about seven or eight degrees. If the axis of the sun be moving, we can see why his nearer and newer planets are nearer the plane of his equator. The more distant and older have had a longer time to depart from his equatorial plane.

3. In the systems of Jupiter and Saturn we see the fulfilment of the same idea. In both cases the distant satellites are farther removed from the equatorial planes of their primaries than those which are near. Here we behold another remarkable coincidence between the general solar and the special planetary systems; and the facts in both cases and the coincidence between the two are explained by our theory. On account of their great distances, nothing is known of the axes of Uranus and Neptune except what is inferred from the orbits of their satellites.

4. It is acknowledged that the great oblateness of Jupiter, 6,800 miles, and of Saturn, 8,060 miles, must have a tendency to hold their nearer satellites in the planes of their equators, that influence not being so potent on their more distant satellites. The case is probably the same with Uranus, and for this reason, as well as from their initial motion in a nebulous state, his moons have followed the movement of his equatorial plane. The nearness of the newer satellites to the planes of their primaries is there-

fore partly due to this influence as well as to their shorter periods for becoming far removed.

5. The small oblateness of the earth, only twenty-six miles, has not had the effect of carrying the plane of the moon's orbit along with that of the earth's equator. And the earth's gravitation on the moon is extremely small; this is indicated by the moon's motion, which is only 2,270 miles an hour; that of the nearest satellite of Jupiter is 38,000, and that of Saturn's nearest satellite is 35,000 miles an hour.

6. The oblateness of the sun is only eighteen miles, according to my own computation. In this respect it nearly coincides with our earth. Correspondingly it coincides with the earth in the fact that the general orbital plane of the planets is far removed from its own equatorial plane, the same as the moon's orbit is far removed from the equatorial plane of the earth. In this point of view the sun and the earth, with very little oblateness, and with their planets and satellites far from their equatorial planes, afford a strong contrast to Jupiter and Saturn, with great oblateness and with the general planes of their satellites near their own equatorial planes. This contrast is significant of the influence of great oblateness on the orbital planes of satellites.

Thus it appears that the apparent irregularities of the solar system, irregularities which at first view seem to oppose our theory, may all be accounted for by that theory, and made foundations for its support. Among the many different bodies of the solar system, the inclinations of their axes to the planes of their orbits, the inclinations of the planetary orbital planes to the plane of the sun's equator, the inclinations of the orbital planes of the satellites to the equatorial planes of their respective primaries, and the retrograde motion of the satellites of Uranus, are all bound up together as effects of the same cause. That cause is the inevitable movement of their axes of rotation. These movements are inconceivably slow, yet we must regard them as absolutely necessary, until we are able to balance a needle by dropping it on its point. If these axes do not move, then they are the only stable things in the known universe.

Besides the motions which the axes of all rotating globes must be supposed to have received in the first act of their formation, there are other influences bearing on them, which have been neglected by astronomers hitherto, because they are infinitesimally small. But these are the very influences now to be taken into view, because we are looking for causes of motions that are infinitesimally small. The motion of the axis of our earth, now pointing to the present north star, and in 12,934 years to point in a direction different by forty-six degrees and fifty-six minutes, is due primarily, strange to say, to the earth's own rotation. By this rotation the ring resulting in the moon was produced, and also its present oblateness. The reaction between this oblateness and the moon causes the slow motion of the earth's axis. Between all other heavenly bodies, including the fixed stars, there must in like manner be reactions in various ways, producing various strange and unexpected effects; but they have been neglected because they are infinitesimally small. But really their united effect is not so small as is generally supposed. The light from a distant star, decreasing all the way inversely as the square of its distance, must be infinitely small, and yet the united light of all these stars enables us to walk out at night with tolerable comfort. How different is our walk by starlight in the fields, from our groping about in a dark room, where no light can enter! Gravity from the same stars also decreases inversely as the squares of their distances, and from each one it is very feeble; but from all together, it may produce effects, strange and unlooked-for effects, during many million years. The influence hourly exerted by the fixed stars upon each other, and upon each other's attendants, may be learned in this way. Let any one calculate the motion in the fixed stars necessary to counteract by centrifugal force the gravity of the sun, and he will be surprised to find how considerable that motion must be. In Alpha Centauri it must be one hundred and forty-five miles an hour. The light of Alpha Centauri is to that of the sun as $2\frac{1}{4}$ to 1, and supposing that its mass is in the same proportion, then our sun must move two hundred and twenty-two miles an hour, to gain centrifugal force enough to counteract the gravity of that star. The light of Sirius is 63.02 times at least greater than that of our sun, and if its mass be in proportion, then our sun must move with a velocity of five hundred and eighty miles an hour, so that its centrifugal force may balance the gravity of Sirius. Thus we find that all the stars visible to the naked eye must be a system, and that there must be velocities among them to counteract each other's gravity; that in these motions and in these mutual reactions, their axes of rotation, in the course of millions of years, may possibly change like that of our earth by the influence of the moon. We do not yet possess the data for estimating these interstellar reactions in all their complicated ways, but we see how they really exist. And from these interplanetary and interstellar influences we infer that motions of the axes of the sun and of the planets may possibly be produced which may account for apparent anomalies. Thus there may be two sets of causes for the motions of the axes of the celestial bodies: the one to be seen in the beginning, in the very formation of the axes, and the other at the present moment, in the interstellar influences.

These motions of the axes and of the orbital planes are

totally different things, and depend on altogether different causes, from those motions of the ecliptic and of the orbital planes of other planets in our solar system, which have formed the subject of such elaborate research by Euler, Lagrange, Laplace, Biot, and others. Let the lines A, B, C, D, and E, represent the orbital planes of some of the planets. As many more may be added as there are planets in our system. Then it is evident that a planet at B would be impelled by gravity toward C, and reciprocally C would be impelled toward B, and both planes would be altered. But in their continual revolutions, some slowly and others rapidly, they would be drawn in contrary directions by other planets, all in their several revolutions acting and reacting on one another, and keeping their several planes in constant motion, each one vibrating to and fro, but evidently within certain limits. This keeping within certain limits by each orbit, is called the stability of the solar system. By their mutual actions and reactions, the position of the general plane of the whole company cannot be permanently changed. It must remain constant so far as their reciprocal influences are concerned, the same as the man remains at rest who attempts to raise himself up by pulling at the object on which he stands. Hence, from the consideration of these forces alone, it has been truly said, that the ecliptic—the earth's orbital plane—never has coincided and never will coincide with the equatorial plane. This is all plain enough. But the nebular theory takes a wider view. It goes upon other data. It looks to interstellar influences. It looks to impulses communicated to nebular masses; impulses arising from irregularities and inquietudes in the nebular condition; impulses comparatively so minute and so far beyond appreciation, that their effects are the mo-

tions of the axes only a degree in many million years. We see the effects of such impulses in the apparent irregularities of the planes and of the axes of the solar system ; and we see how those impulses may have originated in the wild agitation of the nebulæ. The most agitated of all regions known to man is in the surface of the sun. Those agitations were doubtless much larger when the sun was much larger, and when its surface extended to the orbits of the planets.

The unusual inclinations and eccentricities of the orbits of the asteroids probably arose in the beginning from their mutual perturbations. As their orbits are so near, often interlocking, there may have been very near approaches to one another, and these small, light asteroids may have been thrown out of their original orbits into others more eccentric and more inclined. This we know happens to the light comets.

SECTION XXXII.

THE ECCENTRICITIES OF THE PLANETARY ORBITS.

In the ordinary tables, made before the recent determination of the planetary distances from the sun, the perihelion distance of Mercury is put down at 29,000,000 miles, and his aphelion at 44,000,000. The perihelion of Mars is stated at 131,000,000, and his aphelion at 158,000,000. Other planets present similar differences, though generally they are smaller. According to my researches, the action of gravity on a contracting nebulous globe gives a complete account of the origin of these eccentricities. In the nebulous condition the sun was so oblate that the equatorial was about double the polar diameter. Now it is nearly a perfect sphere. According to my calculations, the equatorial exceeds the polar diameter by only eighteen miles; the oblateness being only the $\frac{1}{10000}$ part of the equatorial diameter, which is quite inconsiderable. Newton, in his "Principia," demonstrated how to calculate the oblateness of a rotating fluid. He applied his theorem to the Earth and to Jupiter, and his determinations were wonderfully near the truth, considering the data at his command. Since then, I am inclined to the supposition that the subject has been much neglected by astronomers, and the more so from propositions in some recent numbers of the "Monthly Notices of the Royal Astronomical Society," for making an extensive series of micrometrical measurements to ascertain the oblateness of the sun. If my calculations be correct, the oblateness of the sun is not at all measurable by our micrometry; it is even quite insignificant when compared with the temporary irregularities of the sun's envelope of flame. These irregularities on the contour of the sun are said to be easily visible with good instruments on favorable occasions.

Newton showed that the oblateness of a rotating fluid depends on its density, its size, and its velocity of rotation. His theorem may be applied to the sun and the planets in their former nebulous conditions. Their sizes and densities may easily be found, and their velocities of rotation may be learned from our familiar inclined plane theorem, as well as from the revolutions of the planets and satellites in their orbits. I applied the theorem of Newton to the sun, when its densities, sizes, and velocities corresponded to the orbits and velocities of the several planets, and in this manner found that during nearly its entire nebular history its oblateness remained about the same. Next, I applied his theorem to the planets, when enlarged to their respective satellites, and found the oblateness of all to have been at all times nearly the same as that of the sun. Hence I infer, though I have not the time to work out a demonstration,* that all nebulous bodies condensing and freely rotating under the force of gravity, whatever be their sizes or densities, have the same amount of oblateness. My determinations in all the above cases make the equatorial to the polar diameter nearly as two to one. In the nebulous sun, and in the planets, the effect of the increasing velocity of rotation, naturally causing greater oblateness, was all along counterbalanced by the effect of the increasing density. In them all, however, a great change ultimately occurred. It took place apparently when they were near the act of passing from a gaseous to a liquid condition. This change was caused by the principle of Retardation; a mighty principle, and one which gave some of the most striking and useful characteristics to the solar system. The cause of this retardation in the velocities of the nebulous globes has been fully explained in a former section. With this loss of velocity of rotation there occurred of necessity a loss of oblateness. It must not be understood that the sun and planets rotate with less angular velocity now than they did when they gave off their last rings, but only that their velocities of rotation after giving off their last rings have not kept pace with their amounts of contraction and condensation. It is this want of due and regular increase of rotation that has changed the shapes of all the members of the solar system.

We will now see how this change in the shape of the sun has changed the shape of the orbits of all the planets. Take Mercury, for instance. When the sun was expanded to nearly the orbit of Mercury, and was very oblate, it is evident that his influence on that planet was far otherwise than if he had been a perfect sphere. In a spherical form the influence of his gravity would have been the same as if

* This demonstration, I have since found, was made by Laplace.

all his materials were condensed at his centre. But in the extremely oblate form the influence of his gravity on that planet so situated was considerably greater. Therefore, when the sun contracted, and became nearly a perfect sphere, a great change occurred in the force of gravity acting on that planet. That force became much less. Previously the centrifugal and centripetal forces had been so evenly adjusted, that Mercury revolved in a perfect circle. With the lessening of the centripetal force, what must happen? If, for instance, that lessening were made suddenly, it is clear that the planet would rush somewhat tangentially out of the circle. Its velocity would be too great for the diminished centripetal force. In departing outwardly from the circle, it would go on until its speed was gradually checked by the centripetal force, and then, like a comet, it would be brought back again in an elliptical orbit; and in such an orbit it must forever run, unless affected too much by perturbations from other planets. The change from the oblate to the spherical form of the sun was not indeed made suddenly; but even a slow change in the shape of the sun would produce a corresponding slow change in the shape of the planet's orbit.

What is true of Mercury is true of the other planets. They all began their orbits in perfect circles, because their original ring forms parted evenly all around from the equatorial zone of the sun, which was perfectly circular. When the rings broke, their materials, in globular forms, travelled round in the former paths of the rings. The sun was once in the near vicinity of all the planets, and in the same extremely oblate form. Since then the sun has assumed the nearly spherical form, and his force of gravity on all the planets has become less. Hence, like Mercury, they must all have departed from their perfectly circular orbits, and run into ellipses. Therefore, so far from being an objection to my theory of the action of gravity in the formation of the solar system, the elliptical orbits are an absolute necessity according to that theory.

The only difficulty in the case is to obtain a demonstration how by their material perturbations the eccentricities of the planetary orbits have been modified so very differently from one another. These planetary perturbations are the most perplexing of all the departments of mathematical astronomy. The case of our moon is a good illustration. The lunar theory has been the subject of an immense amount of labor since the time of Newton; and it was perfected only a few years ago. It would have been pronounced to be perfect a hundred and fifty years ago—for even then it seemed complete—but unhappily the open fact was before every observer, that the real motion of the moon was not altogether according to the theory. Some of the data had been overlooked in the calculations, and the trouble was how and where to find them. But when compared with our mundane system, how much more complicated is the problem of the whole solar system! All the planets are continually influencing one another, and there are not only no two months in the year, but there have never been two months in their entire existence, when the influences on any one planet have been exactly the same. How, then, is this difficult and complicated web to be unravelled? Plainly it is impossible. It may be said that some of these influences are so infinitesimally small that they need not be taken into account. They need not indeed for some practical purposes, when an approximation only is all that is required. But when these infinitesimal quantities are going on, and adding up all the time during hundreds of millions of years, they become palpable and great. Nevertheless, it is an established fact that the eccentricities of the planetary orbits may be changed by their mutual perturbations; but the extent of these changes, when the extreme minor influences during millions of years are to be reckoned, is not known, and probably never will be known. Respecting the alterations now going on in the eccentricity of our earth's orbit, there are some disagreements among mathematical astronomers, and well there may be, considering the mazes of difficulties to be encountered. It is cheerfully and thankfully granted that the problem of the stability of the solar system has been solved; but the changes to be effected in all the planetary eccentricities for millions of years is a very different problem, and this has not been solved. The satellites of Jupiter offer a most striking illustration of the influences that celestial bodies may exert on each other. The three nearest have most remarkable relations to one another in their motions, and Laplace has shown that these relations may have in part been caused by their mutual gravitations.

SECTION XXXIII.

ASTEROIDS, METEORITES, AND COMETS.

ASTEROIDS.

The orbits of the asteroids are generally interlinked; that is, they are so near together that the perihelion distance of an outer asteroid is nearer the sun than the aphelion distance of an inner one. They are probably a few hundred in number, about eighty having been discovered in the last twenty years, and they are included within a belt about 150,000,000 miles broad. In view of the dimensions of the rings which formed the planets as given in the thirtieth section, we cannot suppose that a single ring occupied all the space within the asteroid belt. Therefore, the asteroids must have been formed from many rings, of portions of rings; and the same ring, or portion of a ring, may have broken up into several asteroids. This is the more probable, on account of the great disturbing influence of the greatest of all the planets, Jupiter, in their near neighborhood. The eccentricities of their orbits, and the departure of those orbits so far from the general planetary plane, may have been caused by mutual perturbations, especially in their beginnings.

THE METEORITES.

In the rotation of a nebulous globe having the usual extreme oblateness, it must often happen that the centripetal and centrifugal forces are nearly enough balanced to give off a small amount of matter, but not a large ring. The small amounts of matter might not be in the form of continuous unbroken rings. There must have been violent agitations in those ancient nebulous masses, the same as there are still violent agitations in the sun. This unrest was favorable to unconnected separations in small amounts. Even very thin rings, too feeble in their gravitations to collect in a single planet, must have broken up into small fragments, some perhaps when solidified not larger than a pebble. Such small fragments might be parted from the sun, and then they would be of a class with the planets and asteroids; or they might be parted from the planets when in a nebulous state, and then they would be of a class with the moon and the other satellites. On account of their small masses they would be subject to unusual perturbations, and from these extravagant wanderings they might come in collision with the larger celestial bodies, and then there would be a fall of meteorites. Their bursting with loud explosions, their angular forms, and their semi-fused and subsequent semi-vitrified surfaces, would be the necessary consequences of the sudden change from the extreme coldness of space to the high temperature produced by friction through our atmosphere. Their violent velocity is an indication how high that temperature must be. Pouillet and Espy, by entirely different methods, reckon the temperature of space outside of our atmosphere to be two hundred and twenty-four degrees below the zero of Fahrenheit. I believe the temperature of space to be at least a hundred degrees lower than that. The rapid change from so low to so high a temperature would be like throwing a cold stone into the fire. The wonder would be, not the explosion, but the absence of an explosion.

THE COMETS.

The origin of comets, of their motions, and of their nebulous characters, may be easily accounted for by the force of gravity acting on a contracting nebulous mass. It is not necessary that all matter should condense into solids; like the atmospheres of the planets, it may be natural for some bodies always to remain in a gaseous or semi-gaseous condition. The origin of comets may be precisely like that of the meteorites; they may be small portions of matter thrown off in all regions of space, sometimes between the orbits of the satellites, sometimes between the orbits of the planets, and sometimes far beyond the orbits of any planets. Some portions of matter may have remained almost an infinitely long time floating midway between our sun and the nearest fixed stars. After losing their balance, they may begin their wanderings in parabolic and hyperbolic orbits, visiting now one solar system and now another. Some, for instance Lexell's comet, have been turned from their hyperbolic into elliptical orbits around the sun; and in this way they may receive retrograde motions. At least a dozen comets of the shortest periods exhibit strong probabilities of having been produced by the sun since the separation of the ring that formed Saturn. They are always within the orbit of Saturn; they move, like the planets, from west to east; the planes of their orbits are not far from the orbital planes of the planets; and as the orbits of the asteroids are more eccentric than those of the planets, so the orbits of these comets exhibit another and a greater departure from a circle. These eccentricities may be most easily accounted for by perturbations; because we know perfectly well that from time to time the orbits of comets are very greatly changed by proximity to the planets, especially the great planet Jupiter.

The tails of comets come no more within the province of the nebular theory than the aurora borealis of our earth. Both these phenomena belong to other departments of Science. But we must distinguish between the permanently gaseous condition of some comets and the tails of others. These tails are probably temporary, and produced by the nearness of the comets to the sun.

Thus the nebular theory leads to the conclusion that there are small and continuous gradations between planets, satellites, asteroids, meteorites, and comets.

SECTION XXXIV.

DOUBLE, TRIPLE, AND MULTIPLE STARS.—CLUSTERS OF STARS,

REFERENCE has already been made to the different densities of the several layers around the nebular masses. These different densities are the reason why from a rotating nebulous globe the outer materials on the equatorial zone were separated intermittingly as distinct rings, and not continuously as one broad disk. The same fact explains why the nebulous globe forming our mundane system separated into two bodies, the moon and the earth, so nearly of the same size, their diameters being 2,152 and 7,912 miles. Dynamical laws give the same results, whether on a large or a small scale. They have, therefore, made the small system of Jupiter a faithful exemplar of the great solar system. Instead of a small nebulous mass separating into two parts, forming our mundane system, a very large nebulous globe might, by the same laws, separate into two parts, and that would form a double star. Our earth and the moon are in fact a double star. Once they were bright, and shone with an independent light, like the members of the small double star of Gamma Andromedæ, which revolve around their common centre of gravity, and this centre around their primary, the same as our earth and moon revolve around their common centre, and this centre around the sun.

Other stars affords other similar illustrations. These are in truth examples at the same time both of double and triple systems; the same as if our solar system consisted only of the sun, the earth, and the moon. In the same way there might be quadruple and quintuple systems, and also multiple systems consisting of hundreds of stars; as our solar system consists of more than a hundred members. The fact of luminosity or want of luminosity does not touch the dynamical question. Neither does it oppose the nebular theory. On the contrary, these differences in light confirm that theory; for, according to the nebular theory, luminosity is a mere question of time. All celestial bodies must have been luminous once; all will in time be dark. This view of that theory is upheld by current facts; for we know how several stars in modern times have lost their light, the same as our earth, and all the members of the solar system but one.

Double stars, therefore, may be formed by the nebular theory as well as any others. They offer no peculiar difficulties, nor are we in the least led to explain their origin by any peculiar process. As compared with our mundane system, they are mere matters of size; and size causes no difference in the action of dynamical laws. These laws must perform in the same manner, whether the bodies be small, as in the case of the mundane system, or large, as in the case of Castor or Alpha Centauri. Hence it is that already there have been discovered nearly 7,000 double, triple, and multiple stars. Each member of these may have its planets, as in the trapezium of Orion, and each planet again may have its satellites.

It is no argument against the revolving nature of these systems, because in so many cases we can detect no change in their relative positions. Their common proper motions, and the extreme improbability of such juxtapositions by mere chance in perspective, show their physical connections. I have already shown that the differences between the materials of one star and another may be so vastly great as to be comparable to the differences between the several elements of our earth, as platinum and hydrogen. And if some stars are composed of extremely rare materials, then their gravitation must be extremely feeble, and consequently their revolutions must be extremely slow—so slow as to require ages for their detection.

GLOBULAR CLUSTERS.

From what we have learned of our solar system, we cannot suppose that, in a system of multiple stars, they should all revolve in the same plane around their common centre of gravity. We have seen reasons for a departure from a common plane in the cases of the asteroids, the planets, and the satellites. These departures may be so far that the plane of one orbit may be nearly at right angles to another; that is, one may depart, like Pallas in our solar system, thirty-five degrees on one side of the common plane, and another the same distance on the other side, and then to us this would appear to be " a globular cluster." Such a system might be as stable as ours.

We cannot begin to set limits to the numbers of stars that are possible in such clusters. Venus is alone. The system of the earth is double. That of Jupiter is quintuple. That of Saturn, counting his rings, consists of twelve members. The solar system consists of more than a hundred already known, and more are being discovered at the rate of four per annum. Other systems count by thousands, and others again by millions of stars. This brings us by insensible gradations from a single star like Venus to the Nebulæ, and all without varying in the least from the same force of gravity, operating on condensing nebulous masses. In that operation the first step is to make them round; the second is to make them rotate; the third is to separate them by centrifugal and centripetal forces into other clustering systems. These systems are as various as the leaves of the forest, no two being alike. This very diversity is *natural*. And it is a necessary consequence of our theory. It results from uniform laws acting on unlike nebulous masses. Why, it may be asked, were they unlike? For the same reason that all the clouds in our sky are unlike. No two clouds were ever of the same shape, size, and density. From similar principles, when the materials of creation were equally, or nearly equally, diffused through all known space, and then broke up into separate cloud-like nebular masses, these masses must all have been different. Now the great fact that the nebular theory accounts for all this diversity of systems in the heavens—diversity, as far as the telescope can reach—is a powerful confirmation of that theory.

SECTION XXXV.

THE VARIOUS FORMS OF NEBULÆ, OR SIDEREAL SYSTEMS.

WE can understand how a great nebulous sphere might contract without rotation. If perfectly round, and undisturbed from without, all its particles, exterior as well as interior, would settle in radial lines toward the centre. We can understand how a contracting nebulous mass might rotate, and yet not produce a ring. The rotating layer on the exterior might be so thin and light in the beginning, that, as it grew more massive in gaining materials from the unrotating interior, the friction in this process might all along keep the centrifugal a little less powerful than the centripetal force. Then, of course, there could be no ring. Such cases are not only supposable, but they have actually occurred in the interior planets, Mars, Venus, and Mercury, and the asteroids and satellites. We can also understand how a continuous ring might be formed in the shape of a disk; the ring, for instance, beginning to part at Neptune, and continuing to Mercury and even farther, thus forming a flat round plate. Such a production would require a nebulous mass perfectly homogeneous; for if it were made of layers of different densities, and these passing from one to another by sudden transitions, then separate rings would result. Moreover, in the formation of such a disk there must be very little or no retardation. We can understand, also, how in the absence of retardation such a disk might include all the matter of the original nebulous mass; so that after it was perfected from the circumference to the centre, that central part of the disk might not be thicker than the region toward the circumference. If, for instance, the ring separated from the central globe for the distance of two degrees on each side of the equator, these degrees would be greater on the globe when large, and they would grow smaller as the globe grew smaller. Thus the disk would grow thinner toward the centre, but, as a compensation, it would grow denser. Thus every particle or atom of the fluid disk would continue in its own orbit around the common centre of gravity. But we are supposing that chemical action is continuing, and therefore contraction is going on. What, under this continued contraction, would be the fate of the disk? It must contract, but, on account of the centrifugal force, it could not move toward the centre of the disk. Then the result would be this: Just as originally, when matter was equally diffused through all known space, it broke up like a continuous cloudy stratum into separate nebulous masses, so the same thing might be repeated in the stratum of the disk. It might break up into thousands of separate masses, each mass rotating on its own axis, and revolving around the common centre of gravity. This might take place on a very large scale—more easily on a large scale than on a small one; the same as a large planet is more likely to have rings and their resulting satellites than a small one. It might take place on a scale as large as the entire space within the ring of the Milky Way, and even larger. And then the disk, in breaking up into separate masses, each mass forming a solar system, might thus produce millions of fixed stars. Thus, by keeping close to our principles, and working them out on a large scale, we come to results precisely like facts which exist, and which we see scattered generally through space. There are many clusters of stars which, from the faint appearance of their blended light, seem like clouds, and they are therefore called Nebulæ. Many of them are conspicuous for the regularity of their forms, and they are exactly such as may be accounted for on the principles of the nebular theory.

Such a disk as we have just supposed, broken up into many thousands, it may be into many millions, of stars, would be called by astronomers a " Planetary Nebula."

If, instead of an even distribution of the stars in such a disk, there were a crowding together in close proximity toward the centre, giving that centre, at a great distance, almost the appearance of one large star, then such a cluster would be called an " Elliptic Nebula." Such a crowding together toward the centre of the cluster is plainly possible, the same as the planets are in closer proximity in the central region of the solar system. If the planes of such disks were at right angles to the visual ray, they would appear circular; and if they were inclined to that ray in various degrees, they would appear elliptical, with various degrees of eccentricity.

If, in any of these planetary or elliptical nebulæ, many of the individual stars were not to continue their revolutions in the plane of the original disk, but were to depart from that plane by perturbations, some on one side, say thirty-five degrees, like Pallas, and some equally far on the other side, then such an assemblage of stars would appear to us under a globular form, and we would call it a " Globular Nebula." Thus, on the same principles, there may be globular nebulæ on a large scale, as well as globular clusters on a small one.

If, instead of a close proximity of stars toward the centre of the disk-like cluster, there were really one sharply defined, large, bright star, then the cluster would be called a " Nebulous Star." It would correspond to our solar system, with a single great member in the centre; but instead of a hundred attendants formed from rings, it might have hundreds of thousands formed from nebulous masses, into which the original disk broke up. The great central star might be single, or it might be double, and formed like other double stars. Then, of course, the many smaller attendants might revolve around their common centre of gravity.

In the solar system, the largest rings produced were not toward the centre, but far from it. The rings resulting in the four exterior planets were much larger than those forming the four interior. So the outer portion of the disk might possibly absorb by far the largest amount of all the matter of the original nebulous mass. If an outer ring, or outer portion of the disk, were very large, it would attract the matter of the inner portion of the disk toward itself; for that matter, being equally acted upon by the centripetal and centrifugal forces, would receive an outward motion by the gravity of the large outward portion or ring, and thus that ring would increase in size. Afterwards, when by contraction the whole were broken up into stars, that system of stars would present the appearance of a ring, and would be called an "Annular Nebula." The stars inside the ring might be fewer and less crowded than in the ring, and such an assemblage in the heavens might be compared to "a hoop covered with gauze." The action of gravity, in changing a disk into a ring, depends on the truths that a particle anywhere within a spherical shell, and influenced by the gravity of that shell, would remain at rest; but within a ring, unless precisely at the centre, it would move outwardly toward the nearest part of the ring. These theorems are handsomely demonstrated in Herschel's "Outlines," Sec. 735, note.

This last class of objects is like our own sidereal system, consisting of an annulus, or ring of the Milky Way, and a stratum or disk of stars stretching across the whole interior of the ring. Our stellar system, therefore, consists of at least two regions, which we may distinctly describe, although really the two may verge up closely together, and the line of junction cannot be defined. Besides these two regions of our stellar system, there seems to be a third, lying beyond the Milky Way, and consisting of smaller subordinate systems of stars, which, from their vast distance, appear in powerful telescopes like faint patches of light. These outer subordinate systems may hold the same relation to the ring of the Milky Way as the systems of Jupiter, Saturn, and Uranus hold to the ring of the asteroids. But the relative sizes are reversed. These outer systems are probably small when compared with the ring of the Milky Way, whereas the systems of the outer planets are large when compared with the asteroid ring.

The evidence that the stars inside of the Milky Way are disposed in the form of a disk or stratum, appears in the fact, that when we look out at right angles from the plane of the Milky Way, we see comparatively few stars; and as we direct our view more and more toward that ring, the stars become more and more numerous. This is plainly because we look more in the direction of the plane of the disk.

The evidences that the Milky Way is really a ring of stars, appear from the fact that the stars composing that luminous band are generally not far from the same size, thus indicating that their positions are not very far from being equally distant. The telescope pierces entirely through the Milky Way, and then the black ground of infinite space, apparently void, meets the eye. Dark patches, nearly destitute of stars, and in some instances sharply defined, appear in the Milky Way, and these can be easily understood, if we regard that luminous band as a ring; but if we regard it as a stratum, and continuous with the stars nearest to us, then we must suppose that tabular openings, clearly defined, extend off in straight, radiating lines in different directions all the way from our own position through the stratum of stars out to the utmost bounds of their positions; a supposition which no intelligent astronomer will ever make.

The evidences in favor of the third region of our sidereal system are these: In the region of the Milky Way, or close to its borders, appear numerous faint nebulæ which have not been resolved by the telescope into stars. This want of resolvability indicates simply the distance of these great clusters of stars. The question now arises, whether any of these belong to our sidereal system, or whether they are independent systems on a par with our own. They may possibly be the results of rings outside of the Milky Way, concentric with it, and formed originally from the same immense nebulous mass that afterwards gave birth to the Milky Way. Now they are broken up into subordinate systems, analogous to the systems of Jupiter, and Saturn, and Uranus, outside of the ring of the asteroids. There is a possibility of this, and when we have the repeated testimony that these irresolvable nebulæ of a *peculiar character*, distinguished as a class by their irregularities, occur only in the ring of the Milky Way and near its borders, the nebular theory immediately suggests its own peculiar reasons for their position, and gives a probability that they all belong to our sidereal system, and make up its third great region.* The theory of gravity acting on a contracting nebulous mass may thus be generalized and made to account for the existence of all the heavenly bodies and all their different modes of association. By that theory we can see no beginning, except on the supposition that matter was once universally diffused through all known space. The first grand masses into which it separated, formed sidereal systems of various shapes and sizes—planetary, elliptical, globular, annular, and stellar nebulæ. Our

---

* "Outlines," Sec. 869 and 883.

own stellar system belongs to the annular form of nebulæ. It consisted of perhaps several rings, and of a central disk. All the exterior rings are broken into several subordinate stellar groups. The interior and the largest of the rings did not break into different parts so as to lose all traces of its original form, but it separated into stars millions in number, which still occupy the primary position of the ring. The disk interior to the ring also separated into stars, or rather into groups of stars, like our own solar system, the systems of double, triple, and multiple stars, and occasionally into larger clusters like those of the Pleiades, Kappa Crucis, Coma Berenices, Præsepe, and others.

These several divisions and subdivisions can be better understood and remembered when stated numerically, as follows:

The first division was when the original equally diffused material of creation separated into stellar systems.

The second, when, for instance, our stellar system separated into different regions, as the central disk-like stratum, the ring of the Milky Way, and other rings beyond and now broken into minor systems.

The third, when our stratum interior to the Milky Way, and also the Milky Way itself, separated into parts, which afterwards formed solar systems like our own, and also double, triple, and multiple stars, and clusters of stars.

The fourth, when our solar system separated into sun and planets.

The fifth, when the planetary bodies divided and formed satellites.

When we follow out the nebular theory to its logical consequences, we necessarily arrive at a point when all matter was equally diffused, or so nearly equally diffused as to appear such to our conceptions, although an absolute equality of diffusion need not be supposed. That would be unnatural;

for neither in solids, liquids, nor in gases do we ever see an absolute equality of diffusion when on a large scale. This want of absolute equality would account for diversities in shapes, sizes, and densities of the original nebular masses before their rotations began. From these diversities in the original nebulous masses would result, as we have already seen, an infinite diversity of systems and sub-systems.

Of the nebulæ with apparently irregular forms, we need not speak. We do not know that really they are irregular. They may appear such at our great distance for several reasons, even though they may be perfectly regular. It is enough for our theory that so large a number of regular nebulous forms may be accounted for.

SECTION XXXVI.

A CENTRAL SUN?

In this account of the sidereal system to which our own sun belongs, I have said nothing of a central sun. Both the nebular theory and direct observation show that some sidereal systems have large central suns, but that our own system belongs to a class which can have no such great controlling central body. We have already seen, in studying the operations of contracting nebulous masses, that such a mass might contract and solidify without producing a single ring, and hence it could have no attending body. This was the case with the nebulous masses that formed the interior planets, the asteroids, and the satellites. The same might occur on a large scale, providing the nebulous masses were to begin with a nearly round form, and hence with a very shallow rotating current. Then as the current deepened by friction, it might, during the entire period of its contraction, be so much retarded as to keep the centrifugal force always a little weaker than the centripetal force, and hence there could be no ring, and no attending revolving body. From such an extreme case there might be almost an infinite series of sidereal systems, differing from one another by small gradations. The next step in the series might have one attendant like our earth; the next, four like Jupiter; the next, eleven like Saturn; then, a hundred like our sun; then greater and greater numbers, until we come to those sidereal systems which are called nebulous stars, consisting of a great central sun, and perhaps millions of smaller suns revolving around it. But, both by the nebular theory and by observation, the series cannot stop here. The further process is the reduction in size of this central sun, and consequently the comparative increase in the numbers and sizes of the attending suns. The attending suns may, at last, equal the central one, and, from being more closely crowded toward the centre, the system would be called an elliptical nebula. The next step is to have the suns no more closely disposed in the central than in the exterior region of the disk-like stratum, and such a sidereal system would be called a planetary nebula. The next and last term of the series is to have the suns more closely arranged together around the exterior border of the disk, and this would be called an annular nebula. This last is the form of our own sidereal system, and hence it is the farthest of all removed from having a central sun. There may possibly be suns, one after another, from time to time, temporarily occupying in their movements the centre of the system, but such an accidental central sun would not fulfil our idea of a great controlling central sun, holding by its gravitation the other surrounding bodies in their orbits.

The question now arises, What can cause a nebulous mass to take any special place in this series? Why should it condense so as to become altogether a central body, or mostly a ring of stars, or any one of the numerous intermediate forms between these two extremes? If the nebulous masses be very large and very rare, such as are necessary for the production of a sidereal system, then we can see but one cause for their positions in the series, and that is retardation. The greater the retardation, the more central would be the condensation of the mass; the less the retardation, the more quickly would the centrifugal equal the centripetal force, and then the movements toward the centre would be arrested, and the nebulous matter would condense into stars far from the centre, perhaps so far as to form a crowding of stars on the exterior border of the system. A very large external ring already broken up into stars, might so far disturb the equilibrium of these two forces as to draw the stars of the interior disk-like stratum outwardly, and thus increase the numbers in that exterior ring. Thus the principle of retardation, which, as we have seen, had so much to do in giving the present characteristics to our solar system, must, by the same dynamical principles, have had an equal agency in giving their characteristics to other systems, to clusters of stars, and to the various forms of nebulæ, which are so many various sidereal systems. The amount of retardation which a contracting mass receives must depend mostly on its original shape. If compact in form and nearly round, the retardation would be great; if very irregular, the retardation would be small. The more scattered and irregular, the more the interior would be put in motion in assuming the spherical form, and hence it would not receive its rotatory motion alone by friction from the exterior layer, and hence, also, it would not retard that exterior. This, however, is on the supposition that the motions of the exterior and the interior are in the same direction, which, in general, they would most likely be.

We cannot say that on dynamical principles a central sun is a necessity in order to secure stability and permanence. A great system of bodies, for aught we can see, may revolve with just as much stability around their common centre of gravity as around a powerfully attractive central body. To my mind the stability of a system seems far more secure when the revolutions are around a common centre of gravity which is absolutely vacant space. For then there can be no collision with such a centre.

Moreover, simple observation is as strong as any argument can be, to prove that our sidereal system has no central sun. If it existed, it would be luminous; for, like our sun, the central body, being larger, must retain its luminosity longer than its attendants. But we see no such large central body in our system, although we are situated very nearly where it ought to be. The facts which prove that we are not far from the centre of our sidereal system are these: First, we are near the plane of the Milky Way, because that ring apparently divides all space into two nearly equal parts. Secondly, we are nearly equally distant from every part of that great ring, because it appears on every side nearly equally bright. Glowing descriptions have been given of its greater brightness in the southern hemisphere, but these convey rather the surprise of travellers, the expression of their emotions, at a greater degree of brightness, than an accurate scientific comparison, and therefore they are apt to mislead. My own observations were frequent and careful during more than three years, and my recollections are very vivid, although a quarter of a century has elapsed. The general effulgence of the heavens in some southern regions is very strong, but in my opinion the effect on the mind arises from the assemblages of large stars such as the Southern Cross, and conspicuous nebulæ, such as the Magellanic Clouds, as much as from the superior brightness of the Milky Way. Still, on account of that superior brightness, we are a little nearer to that great ring in the southern direction. Thirdly, we are not far from the central plane of the stratum of stars inside the Milky Way, because in counting the stars within equal degrees of nearness to both poles of the Milky Way, their numbers are not far from being equal; whereas, if we were on or near one surface of that stratum, the stars would be exceedingly more numerous in the direction of the opposite surface. While, therefore, we are not exactly in the centre of our sidereal system, we are nearly there; and if a great central sun existed with a size proportionate to the system, we would be dazzled and burned by its rays. The motion around it of our solar system would be rapid beyond all comparison, and we would quickly see corresponding real and apparent motions of the fixed stars. Their real motions would be enormous, to counteract the centripetal force of so great a central sun.

SECTION XXXVII.

THE PROPER MOTIONS OF THE FIXED STARS.

The forms of the various sidereal systems may fairly be regarded as inevitable consequences of gravity acting on contracting nebulous masses. We have seen how nebulous stars, elliptical nebulæ, globular nebulæ, planetary nebulæ, and annular nebulæ are just such forms as must result from contracting nebulous masses under the influence of gravity. The forms of those several clusters must depend on the amount of retardation; and the amount of retardation must depend on the original shape of the nebulous masses before the spherical form was attained. If compact and nearly round, retardation must have been great; if very irreg ular, scattered, and straggling, then retardation would be slight.

All the evidences which prove that the nebulæ received their regular forms from the action of gravity, prove also that the proper motions of the stars were caused by gravity. In our solar system the same force gave both form and motion.

The proper motions of the fixed stars are not a mere theoretical assumption. It is a well-grounded fact. These motions must have been derived from some powerful cause. Gravity we have found to be such a cause. It has given motion to the stars of our solar system; and these stars, although from their relation to the sun they are called planets, must nevertheless be regarded as of the same nature as suns and fixed stars. The reality of their being fixed stars we saw well established in the beginning of this volume, and now we see another of its logical consequences. The same cause which formerly lighted up our earth in a blaze like the sun, must also have lighted up the other stars; and the same cause which gave its rapid motion to our earth, must in like manner have given their motions to the other stars.

The grounds of our belief that gravity was the cause of the motions of the fixed stars are therefore more than one: first, gravity has caused the motions of some stars—the planets, which are of the same general class as the fixed stars; secondly, gravity has given those distant sidereal systems their various regular forms, and therefore it has given them their motions. Another ground for this belief is seen in the fact, that we know of no other physical cause adequate for producing such motions, and we can conceive of no other physical cause.

One of the greatest unsolved problems in astronomy in our era is the combined system of movements of the stars of our own sidereal system. It is worthy of the application of the highest talents and of the largest learning, and it seems to invite especially the devotion of young men who may make it their life-long labor. There are at least four aids and encouragements to undertake the work: first, the form of our sidereal system, which is a ring with an interior disk-like stratum, and both the ring and the stratum lie in the same plane; secondly, our position in the system, which, as already shown, is near the centre; thirdly, the origin of the system and its motions, according to the nebular theory; fourthly, the present calculable influence of the stars on one another, such particularly whose distances are known.

The aid we may receive in the solution of this great problem, from our position near the centre of our sidereal system, may be estimated by our relation to our solar system. How much more simple would the motions of the planets appear from a central point, like the sun, than from a half-way position, as on our earth! Knowing, therefore, our stand point in our sidereal system, we may put together the motions of the fixed stars with the more confidence and hope of final success in unravelling the great mystery of their intricate dance.

The first hint which the nebular theory supplies is, that the general motions of the fixed stars of our system are in or near the plane of the Milky Way. The second is, that generally these motions are in the same direction. The third is, that the orbits of some stars may be much inclined to the plane of the Milky Way; and that two orbits may be inclined in opposite directions, and hence their motions may be nearly at right angles to each other. The fourth is, that there are subordinate systems in the general stellar system; the same as there are planetary systems within the solar system. A few stars which partake of the common direct motion in the plane of the Milky Way, may appear to us to have a contrary motion, the same as some planetary motions seem retrograde.

Vast as this problem of stellar motion may appear, we are not certain that it is really more difficult than that which was solved by Copernicus. But the number of objects is far greater, and their apparently slow motions may require vast labor and a long time.

Much assistance may be derived from the facts that the fixed stars are at this moment exercising a powerful influence on one another through the force of gravity, and that when their distances are known this influence is calculable. Recently, after computing the amount of this influence, I was astonished at its greatness and its importance in a scientific point of view. I found that the force of the sun's gravity on Alpha Centauri is such, that to avoid falling into the sun, that star must move at right angles to its present direction at the rate of one hundred and forty-five miles per hour. The light of Alpha Centauri is estimated to be 2.32 times greater than the sun's, and if his mass holds the same proportion, then the sun must move two hundred and twenty-two miles an hour at right angles to the direction of Alpha Centauri, to avoid falling into that star. This is on the supposition that the sun is acted on by the mass of Alpha Centauri alone. Sirius has a light at least sixty-three times greater than that of the sun. One computation of good authority, that of Clark, of Boston, makes its light more than two hundred times greater than the sun's. But if the mass of Sirius exceeds the sun's only sixty-three times, then, on the supposition that the sun is acted on by the mass of Sirius alone, it must travel at the rate of five hundred and eighty miles an hour in or near a circle around that great star, to gain a centrifugal force equal to its centripetal force. Two methods for calculating these results are given in Appendix III. These I regard as very great facts in astronomy. They shed new light on the intimate structure of our stellar system. They show that all its visible stars are woven in a single web. They show that at least a hundred suns are acting strongly, and others more feebly, on our own sun and his revolving attendants. At the same time, our own acts strongly on a hundred others that are more near, and feebly on a greater number that are farther off. While each star is thus impelled by many opposing influences on all sides, and while these influences are, some stronger and some weaker, there must result a very complicated system of movements among them all, besides the general drift of their entire movements in the plane of the Milky Way.

Just as the sun acts on the mundane system, the earth and the moon, and causes in that system various irregularities, and as the moon acts on the earth and causes in a few thousand years a great change in the direction of its axis of rotation, so we may be sure the fixed stars produce changes in the solar system, and hence to their power, in part at least, we may attribute changes in the inclinations of the axes of rotation. Hence we need not wonder if the general plane of the planetary system is not exactly in the plane of the sun's equator. These changes may be too small and too slow to admit of calculation, yet, in the course of millions or hundreds of millions of years, they may amount to a few degrees, and thus become strikingly apparent. Even the shapes, sizes, and positions of the planetary orbits may thus be somewhat modified.

When we know the distances of two stars, and the angle of divergence between their directions, then we may know their distances from each other. And then, on the assumption that their masses are as their light, we may calculate their mutual influences on each other by the force of gravity.

In this way, by their known proper motions, by their known distances, by triangulation, and by calculation, there is a promise of discoveries, not only in the particular masses and movements of the fixed stars, but also in their general drift or combination in the disk within the Milky Way.

PART IV.

THE CAUSE OF THE MOTION OF THE STARS.

SECTION XXXVIII.

GRAVITY IS THE CAUSE OF THE MOTIONS OF THE STARS.

The movements of the stars along the blue vault of heaven have from time immemorial been a wonder and a subject of admiration both to the savage and to the sage. Doubtless these motions first led to the study of astronomy. All seem to proceed night after night with perfect uniformity from east to west. And this moving together, as with a single turn of a wheel, gave origin to the name Universe, which means one turn. A few stars seemed irregular in their motions, and hence they were called planets or wanderers. Even the sun seemed to change his place relative to the stars, and to make an entire circle around them once every year. Hence the chief labor of astronomers, not only in the early and in the middle ages, but also in modern times, has been to learn the exact motions of the stars; and probably none has received greater attention than our moon.

It is a remarkable fact that, while the movements of the stars have been studied with so much care, the cause of these movements has so long been hidden. Even a conjecture has scarcely been made of the cause of their velocities. Some one of antiquity said that if the stars had fallen toward the earth, and had been turned at right angles from their former courses into their present apparent directions from east to west, their velocities would have been the same as now. But this must always have seemed as an idle dream. No cause appeared for turning them at right angles to the directions of their fall, and no possible estimate could have been made of the velocities of such fall. In modern times we know that these movements are not real; they are only apparent from the diurnal rotation of the earth. And those motions which are real are not from east to west, but in the contrary direction, from west to east.

Since the reception of the Copernican theory, and the true knowledge of the motions of the stars, mankind have been filled with awe at the astonishing velocities of those great bodies. Sixty-eight thousand miles an hour is inconceivable even for a small body; but when we learn that the vast and ponderous globe on which we live, flies along with such lightning speed, we are awed from inquiring what could have imparted such an impulse, and we are disposed to refer the cause to the direct and miraculous interposition of God. Hence, until now, the true cause of the motions of the stars has not been discovered, and not even conjectured, as far as I am aware. Gravity is known to be a great motive power, but there has been a singular unanimity among scientific men in declaring that gravity could not have produced the motions of the stars. This I will soon show by quotations from many authors, beginning with Newton, and coming down to the very latest writers. Laplace, and many of his followers, rightly believed that the solar system originated from a rotating nebulous envelope around the sun; but they received that rotation as an existing fact, without inquiring into its cause. A few have supposed that the rotation was caused by the particles of a nebulous mass condensing and meeting at the centre, but they all failed in finding a force to produce the rotation. Moreover, I will soon show that no rotation could possibly be produced at the centre of such a mass. In that region, and for millions of miles around in every direction, all was still, or nearly still, and motionless. Rotation must necessarily begin on the surface, and proceed gradually toward the centre, the very last point, instead of the first, to rotate.

A few years have elapsed since I first began, in conversations, to state my belief that gravity was the origin of the motions of the stars. The objections I have met may here be stated together, with the considerations by which these objections are overcome. It has been said that "the relation between the orbital velocities of the planets, and the velocities of falls produced by gravity, proves nothing as to the source of their orbital velocities. That relation proves only that those orbital velocities are *compatible* with the assumption that they are due to gravity. They may possibly have been produced by some other physical cause. And they may have been produced by a miraculous act of the Almighty." The following facts completely remove these objections:

1. Gravity is a power strong enough to produce these motions. This becomes evident by the inclined plane theorem. No phenomena in the universe more vividly impress the mind with the idea of omnipotence than the force necessary to send a great world like our own through space with a velocity of 68,000 miles per hour! In gravity we have found a force competent for this momentum, and not only for this, but for overcoming a certain amount of friction which must occur in a contracting and rotating nebulous globe.

2. Gravity varies its power. Near the sun, it is strong; far off, it is weak. Just in proportion to these variations in force, are the variations in the velocities of the heavenly bodies. Gravity could not produce the velocity of Mercury away off at Neptune, and accordingly we see no such velocities there. Through this wide distance we see everywhere a correspondence between the power and the velocity as between cause and effect. In other words, the velocities of all these one hundred and fifteen bodies of our solar system correspond with the heights of the inclined planes down which they have come, and to the powers at work on those planes. The planes were all of different heights, and the motive powers on each one were different, and in all these one hundred and fifteen cases the velocities are different and coincident with the planes and powers.

3. The motions of these one hundred and fifteen bodies are varied in their directions, and complicated. All have two motions, and the satellites have three. Could we see with one view of the eye a piece of mechanism composed of one hundred and fifteen parts, and all these parts moving together in double and triple motions in exact accordance with those of the solar system, we would be filled with wonder. Gravity, acting on a contracting nebulous mass, is a force exactly fitted to give origin to all this complication of motion!

4. In satisfying ourselves of the connection between an effect and a cause, we first ascertain whether they can be in each other's presence. It is not sufficient to know that a certain force is great enough to produce a given effect: we must know also whether they can exist together either in actual contact or through some medium. In the solar system, where we find these great, these various, and these complicated motions, just there also we find gravity in actual presence as an appropriate cause.

5. There is no other known physical cause for these motions. We can think of no other physical cause. Gravity is in every way fitted for these strange effects. Therefore it is the true cause. This is the way that all men, learned and unlearned, reason on every other subject, and we must reason the same way here. After finding a cause competent for a given effect; after examining all the peculiarities in the effect, and finding the same corresponding peculiarities in the cause; after seeing the cause and the effect answering to each other like a stamp answering to a seal, we must conclude that they belong to each other. This is the case between the strangely complicated phenomena of the solar system on the one hand, and on the other the force of gravity acting on a contracting nebulous mass.

6. But it is not alone the various velocities of more than a hundred bodies which coincide with the force of gravity. It is not alone the *directions* of these velocities in complicated whirls, like a mazy dance, which coincide with that all-prevailing force. In addition to all these, there stands out the great fact that the formation of the entire mechanism of the solar system, the separation of all the parts one from another, the placing of these parts in their present orbits, the shaping of these parts in their present forms, all coincide with the action of gravity. Here, in the formation of the solar system, is a great number of facts; and the greater the number of facts which coincide with any theory, the stronger is the support they lend to that theory. When, as I have already repeated, a lock has many intricate interior guards, and a key is found precisely to fit all those guards, and to move the bolt easily backward and forward, then we know we have the true key. It is so in the problem before us. Therefore it may be well to review the many facts in the formation of the solar system, and see how very numerous are the correspondences between that system

FACTS ACCOUNTED FOR BY GRAVITY. 321

and the inevitable action of gravity on a contracting nebulous mass.

SECTION XXXIX.

A SUMMARY OF FACTS IN THE SOLAR SYSTEM ACCOUNTED FOR BY GRAVITY.

IN the beginning of this discussion on the origin of the stars and the cause of their motions, I proposed to show that the solar system, with all its motions and all its wonderful peculiarities, may be accounted for by the action of the force of gravity on a condensing nebulous mass; or, in other words, that the solar system is an inevitable consequence of such action. The following is a summary of the facts so accounted for in the preceding pages:

1. Gravity accounts for the round forms of the stars.
2. It accounts for the rotations of the stars.
3. It accounts for their oblateness; because the inertia or centrifugal force producing that oblateness is merely the force of gravity conserved.
4. It accounts for the separation of the satellites from the planets, and of the planets from the sun; or, in other words, it accounts for the independent existence of all these one hundred and fifteen bodies.
5. It accounts for the existence of the unbroken rings of Saturn.
6. It accounts for the broken ring of the asteroids.
7. It accounts for the approach to regularity in the planetary distances from the sun.
8. It accounts for the velocities of the several planets, asteroids, and satellites.
9. It shows why the velocity of the sun's equator is slower than the orbital velocities of the planets.
10. It shows why the equatorial velocities of Saturn

14*

and Jupiter are less than the orbital velocities of their inner satellites.

11. It shows why the equatorial velocity of the earth is slower than the orbital velocity of the moon.

12. It shows why Jupiter and Saturn, contrary to the case of the sun, have greater equatorial velocities than the orbital velocities of their farthest attendants. Those two planets suffered less retardation than the sun.

13. It shows why the four exterior planets have all the satellites but one.

14. It shows why, of the four interior planets, the earth alone has a satellite.

15. It shows why Saturn's exterior satellite is double the distance of any other satellite from its primary.

16. It accounts for the nearness of Saturn's rings to that planet.

17. It shows why Saturn has so many more attendants than any other planet.

18. It shows why the satellites are so much nearer together than the planets.

19. It accounts for the fact that there is no planet interior to Mercury.

20. It accounts for the wider interplanetary space between Mercury and Venus than between Venus and the Earth.

21. It accounts for the wider intersatellite space between the first and second satellites of Saturn and Uranus than between the second and third, counting from within outward.

22. It accounts for the fact that the velocities of the rings of Saturn around the sun are precisely equal to that of Saturn.

23. It shows how the satellites acquired their three simultaneous motions!

FACTS ACCOUNTED FOR BY GRAVITY. 323

24. It accounts for the comparatively large distance of the earth's satellite, and for its slow motion.

25. It accounts for the large size, relative to the earth, of the earth's satellite.

26. It accounts for the revolutions of the planets around the sun nearly in the same plane.

27. It accounts for the revolutions of the planets around the sun in the same direction.

28. It accounts for the revolutions of the planets in the same direction as the sun's rotation.

29. It accounts for the revolutions of the satellites around the planets, nearly in the same plane, in the same direction, and that direction the same as the rotation of the planets, and nearly in the equatorial planes of the planets.

30. It accounts for the rotations both of the planets and the satellites from west to east.

31. It accounts for the fact that the satellites make only one rotation in one revolution.

32. It accounts for the law of density of the solar system; that is, why the inner planets are more dense than the exterior ones, why the planets are more dense than their satellites, and why the interiors of the planets are more dense than their exteriors.

33. It accounts for the elliptic forms of the planetary orbits, and for the near approach of those ellipses to circles.

34. It accounts for the inclination of the earth's axis of rotation to its orbital plane, thereby causing the seasons, and also for the same facts in the other planets.

35. It shows why the orbital planes of the planets are not precisely in the equatorial plane of the sun.

36. It shows why the orbital planes of the satellites are not in all cases precisely in the equatorial planes of the planets.

37. It shows why the nearest planets are nearest to the sun's equatorial plane.

38. It shows why the nearest satellites are nearest the equatorial planes of their primaries.

39. It accounts for the retrograde motions of the moons of Uranus.

40. It accounts for the origin of meteorites.

41. It accounts for the origin of comets.

42. It accounts for the systems of double, triple, and multiple stars.

43. It accounts for clusters of stars.

44. It accounts for nebulous stars.

45. It accounts for elliptical nebulæ.

46. It accounts for planetary nebulæ.

47. It accounts for annular nebulæ.

48. It accounts for the ring of the Milky Way.

49. It accounts for the disk-like form of the collection of stars interior to the Milky Way.

50. It accounts for the fact that the plane of this disk-like stratum is in the plane of the Milky Way.

51. It accounts for the many irregular nebulæ in or near the Milky Way.

52. It accounts for the proper motion of the so-called fixed stars.

53. It accounts for the fact that the planetary systems, consisting of planets and satellites, are miniature resemblances of the solar system.

This is an extraordinary catalogue of facts—extraordinary for their greatness, for their differences one from another, and for their number. Even those which seem most familiar, are truly mighty in bearing conviction. For instance, the common one of rotation: every heavenly body rotates, and this is a necessary consequence of the action of gravity. We cannot conceive how an irregular contracting nebulous mass, under the influence of gravity, can exist without rotating. The want of such rotation is an impossibility. Many of these facts are marvellously strange; such as the three simultaneous motions of the satellites, and the motions of Saturn's rings around the sun, precisely even with the motion of Saturn. And yet how perfectly and beautifully does the action of gravity account for them all! How many motions there are both of rotation and revolution from west to east in the sun, the planets, the asteroids, and the satellites! How remarkably do they all approach the same general plane! And how naturally all these are accounted for by the influence of one force! Even the inclinations of the axes of rotation to the planes of the orbits, and the inclination of Uranus going so far as to produce retrograde motion, harmonize with this theory. How different is the system of Saturn from that of Jupiter! Saturn, although much the smaller star, lacks but one of having three times as many attendants; the farthest being twice the distance of Jupiter's, and the nearest being five times more near. Jupiter's rings are all broken, and three of Saturn's still remain. Yet these differences are accounted for clearly and simply by the less retardation which gravity encountered in giving rotation to Saturn. How easily does this theory account for the facts that the four exterior planets have all the satellites but one, and that of the four interior planets the earth alone has a satellite! We need recapitulate no further. If ever a theory was established firmly, triumphantly, and gloriously by facts, then is this theory of the action of gravity so established.

SECTION XL.

A COMPARISON BETWEEN THE EVIDENCES THAT GRAVITY NOW HOLDS THE PLANETS IN THEIR ORBITS, AND THE EVIDENCES THAT GRAVITY ORIGINALLY GAVE THEM THEIR MOTIONS IN THEIR ORBITS.

The evidences are as strong that gravity originally gave the stars their motions in their orbits as that now it holds them in their orbits. To make this truth plain, let us for a moment examine the process by which Newton discovered this great truth.

1. By a very simple arithmetical calculation, founded on the distance and the velocity of the moon, he learned the amount of the centrifugal force of the moon. By another equally simple arithmetical calculation, founded on the amount of force in the earth's gravity at its surface, or 8,000 miles from its centre, and on the theory that this gravity decreases as the squares of the distances increase, he learned what would be the amount of the earth's gravity at the distance of the moon. Thus he found that these two forces, the centrifugal of the moon, and the earth's gravity acting on the moon, or the centripetal force, balanced or nearly balanced each other. This was his grand discovery!

2. Next, by a like calculation, founded on the velocity of the earth, and on its distance from the sun, he ascertained the centrifugal force of the earth. He then *assumed* that the mass of the sun was great enough by the force of its gravity to hold the earth in its orbit; that is, he *assumed* a centripetal force in the sun acting on the earth, equal to the centrifugal force of the earth. Taking this assumed mass of the sun or its gravity as a basis, he applied it to the other planets, and found by calculation that, at their respective distances, the gravity toward the sun, or the centripetal

force, equalled or nearly equalled their several centrifugal forces. Here was a new confirmation of his discovery!

In both these cases it was clear to him that there must be some force to counteract the centrifugal force of the moon and also of the planets. Gravity, if extended to these bodies, would just counterbalance these forces. There was no other known physical force to do this counterbalancing. Gravity was a sufficient force. Therefore, gravity he believed to be the true force. This is precisely the same as our reasoning when we concluded that gravity gave the moon and the planets their motions in their orbits. The very language is identical with what we employed only a few pages distant. We said that a certain amount of force had been necessary to give the planets and satellites their respective velocities in their orbits. The force of gravity was just that amount, and also enough for the small but necessary friction. Moreover, it was known to have been right in the presence of these planets, and acting by theory so as to produce not only their velocities, but also their very *directions!* No other known physical force could so act, either as to power or direction. Therefore we were compelled to conclude that gravity produced these motions.

If we look at the philosophy of this subject, we will perceive that the ground of Newton's belief was the coincidence between the theory and the facts. The theory was, that gravity impels the moon toward the earth with a precise amount of force. The fact is, that the moon is actually impelled toward the earth with that same amount of force. This coincidence created the belief that gravity is the real impelling force, especially as no other force is known. The very many coincidences between theory and fact respecting the origin of the motions of the stars, must now have a like effect in creating in our minds the belief that gravity produced these motions.

3. But these were far, very far from being all the coincidences or evidences which Newton had to prove, that the stars were held in their orbits by gravity. He proved that very many strange phenomena in their motions, could be accounted for by gravity, and by gravity only. Among these were the precession of the equinoxes; the movements of the stars, not in circles, but in ellipses; the situation of the sun in one of the foci of those ellipses; the equal spaces passed over by the radii vectores in equal times; the proportions between the velocities and distances of the several planets, and their mutual perturbations and irregularities. All these, and many other phenomena, he proved by most difficult mathematical reasonings to be the inevitable consequences of the action of gravity. That is, these very strange motions of the stars are just such as they would be if gravity were acting upon them. He could think of no other force to produce these actions—therefore he believed they were produced by gravity.

In like manner, the coincidences between the velocities of the planets and the motive-power of gravity are far, very far, from being all the evidences which prove that gravity gave their motions to the stars. There are very many other strange phenomena in the solar and stellar systems, the origin of which can be accounted for by gravity, and by gravity only. Among these we may point to the long catalogue in the last section, a most triumphant assemblage of great facts, all completely *coincident* with our theory; all showing themselves to be the inevitable consequences of the action of gravity.

If we make a careful enumeration of all the coincidences between theory and fact, which prove that gravity originally gave their motions to the stars, we will find that they are more numerous than the coincidences between theory and fact, which prove that gravity now holds them in their orbits. For instance, the very large class of rotations in the sun, and in his many attendants, are coincidences to prove the origin of stellar motion; but there is nothing analogous to this to prove the cause of the centripetal force.

The mathematical demonstrations in both cases are strict and conclusive; and if, in the one case, a small excess of the force of gravity is accounted for as having been expended on a necessary friction down an inclined plane, so in the other case there are slight discrepancies in numbers, but not enough to destroy the rational and satisfactory coincidences between theory and fact.

In Newton's day it seemed incredible that gravity should reach so far—should extend through all space. In our own day it seems incredible to many that matter should have been expanded so far—should have been extended through all known space. But the proofs of these extensions in both cases are the same; numerical calculations show that these two theories agree perfectly with existing facts. Even now we know that matter is diffused completely through all known space. There is no vacuum. The fluids whose vibrations are called light and heat, and which travel with such inconceivable velocity, are now extended everywhere. The same may be said of the material medium which produces gravity. They are indeed extremely rare, yet we cannot say they are more rare than the materials of the solar system when expanded a thousand times farther than the orbit of Neptune.

The matter of the solar system is now arranged exactly as it must necessarily be arranged on the theory that it was formerly so expanded. While the chief mass is concentrated in the rotating sun, yet far away from his surface we see it strewed here and there along the path by which it contracted. Mercury is more than 30,000,000 miles from the sun, and thence outwardly we see the materials scattered at intervals until Neptune, which is a hundred times farther off than Mercury. Then, if we look away to other solar systems, we see other small stars revolving planet-like around great central orbs, the same as here. There, the same as here, "the nearest companions have motions ten times more rapid than the farthest ones." If, from solar we look still farther to stellar systems, we see immense suns arranged in clusters called nebulous stars, elliptical nebulæ, planetary nebulæ, and annular nebulæ; among these last is ranked our own Milky Way, with its interior disk-like stratum of stars. In all these regular forms of nebulæ, and in the system of double, triple, and multiple stars and clusters of stars, we see matter arranged and moved just exactly as it must necessarily be arranged and moved on the theory that it once was universally expanded in a nebulous condition, and afterwards contracted and impelled by the force of gravity. Those regular nebulous forms, and the systems of double, triple, and multiple stars, add powerfully to the coincidences between theory and fact which we find in our small solar system.

The parallelism between the evidences that gravity gave the stars their motions in their orbits, and that gravity now holds them in our orbits, is seen still further in the fact that gravity stands in the same relation to the Nebular Theory as it does to the Copernican Theory. In the nebular theory, gravity imparts motion to the stars; in the Copernican theory it keeps the stars in their courses. In the one theory it originates motion; in the other it regulates motion. By keeping this idea in view, we can the more easily compare the evidences in the two cases.

Thus far the work of gravity may appear equally sublime, equally soul-elevating in both these theories. But in the nebular theory it does far more than merely originate motion; it performs a far more sublime and more soul-ennobling part. Acting in its direct way, and by its conservation in the form of inertia, it separates one body from another, shapes them, rotates them, arranges them in their orbits, and there, in those orbits, it imparts their astonishing projectile velocities! A philosopher in a former age, while watching insect transformations—the egg, the caterpillar, the chrysalis, and the butterfly—exclaimed, "Here is seen the hand of God!" May we not give expression to our emotions in the same words, while tracing the operations of gravity?

SECTION XLI.

THE OPINIONS OF SCIENTIFIC MEN HITHERTO ON THE CAUSE OF THE MOTIONS OF THE STARS.

THE inquiries will naturally arise, How has it happened that gravity has never before been discovered to be the cause of the motions of the stars? And what have been the speculations and opinions of scientific men on this subject? These inquiries will best be answered by perusing their own writings; and I have put in Appendix IV. a list of all the authors on this subject that I know, with special reference to their works. They number nineteen in all, and doubtless there are many more, that I do not know. Here I will give a short but true sketch of what each one believed:

NEWTON.—To the venerable Sir Isaac Newton is due the first place in the catalogue. His language is as follows: "But yet I must profess I know no sufficient natural cause of the earth's diurnal rotation."

"The planets and comets will constantly pursue their revolutions in orbits given in kind and position, according to laws above explained. But though these bodies may indeed persevere in their orbits by the mere *laws of gravity*, yet *they could by no means have at first derived the regular position of the orbits themselves from these laws.*"

After describing the orbits, and positions of the planets and satellites, he says : " But it is not to be conceived that mere mechanical causes could give birth to so many regular motions." By " mechanical causes," he means those which act on a body from without, and among these he puts gravity. He evidently could think of no other cause for the motions of the stars but the immediate interposition of the Great First Cause of all things. He says : " This most beautiful system of sun, planets, and satellites could proceed only from the counsel and dominion of an intelligent and powerful Being." And immediately he goes on at some length to give his ideas of this Being.

HERSCHEL, the elder.—This truly great man speculated not on the formation of the solar system, or on the separation of a nebulous globe into planets and satellites, but simply on the condensation of diffused nebulous matter into stars, and on the cause of their rotations. His first papers were presented to a Philosophical Society at Bath, in 1780 and 1781. Papers also touching on the cause of rotation he published in the Royal Philosophical Society in 1789 and in 1811. The origin of rotation, therefore, occupied his thoughts more than thirty years at least ; and between his first and last papers his ideas considerably improved. He saw many nebulæ with round forms ; to attain and to support those forms, he believed rotation to be necessary ; and the causes of this rotation he supposed might be gravity and " other central forces," to us, as yet, unknown. Other subsequent authors also, as we shall soon see, have believed in unknown mysterious forces as being necessary to produce rotation. At length, in 1811, he thought gravity alone might be sufficient. But he put forth his opinions in queries and as "a surmise," and without a conclusive demonstration. His chief merit in this regard was to assign the prominences or "eccentric matter" of a nebulous mass as the causes of rotation. "These, in displacing other matter, or in sliding down sideways, must produce a circular motion." He did not, however, have distinct ideas: 1st. That the sliding down sideways of this eccentric matter would produce simply many currents, in various directions, acting against one another on the surface of a nebulous globe. 2d. That these currents must ultimately all result into one. 3d. That this one current could not possibly stop, but must, of necessity, go on faster and faster every moment, in consequence of the continued action of gravity, and the contraction of the nebulous globe. From the vagueness of his surmises, they have not been made the foundation of further progress by other authors. They have been quoted by only one scientific man, that I am aware, Professor Heinrichs; and he, not seeing the conclusiveness of Sir William's statements, devised another and very different theory of rotation. The visible nebulæ, however, on which Herschel speculated, so far from being diffused, cloud-like, nebulous matter, are now known to be clusters of stars, sidereal systems, so very far away, that their united light seems faint and dim. Moreover, he had no idea of the formation of planets from a rotating nebulous mass, and hence said not a word about gravity as the cause of the projectile or proper motions of a star.

LAPLACE.—The celebrated theory of Laplace must not be regarded historically as a continuation or a development of the nebular speculations of the elder Herschel. Herschel wrote on the condensation of nebulous matter into fixed stars and suns. Laplace wrote on the formation of the planets and satellites of our own system, from an atmosphere of the sun. He assumed the sun as already formed, and endowed with its present rate of rotation. Next, he supposed that, from some extraordinary cause, the sun's heat and light were vastly increased, like the temporary star which appeared in 1572, and shone with such transcendent brilliancy. From this temporary heat, he supposed the atmosphere of the sun was expanded beyond the orbits of any of the planets, and in cooling and contracting, its equatorial zone parted several rings of nebulous matter, which subsided into nebulous planets, and from these were derived the satellites. He says not a word about the cause of the rotation of the sun or of his attendants; and he names Buffon, not Herschel, as his only predecessor in speculating on the same subject. Sir John Herschel, in his "Outlines," declares that "the nebular hypothesis" of Laplace, and "the theory of sidereal aggregation" of his father, "stand, in fact, quite independent of each other."

BUFFON.—The speculations of Buffon I have not read. Laplace carefully describes and refutes his theory, which is, that the planets, and consequently their motions, were derived from a collision between a comet and the sun.

KANT.—This philosopher is said to have preceded Laplace many years in forming a similar theory. I have not access to that part of his works; but one of his advocates, Professor Heinrichs, says that, like Laplace, he assumed the rotation of the primitive body, without accounting for the cause of that rotation.

MARY SOMMERVILLE.—This highly intellectual lady, writing of "the primitive cause which determined the planetary motions," says, that "Laplace has computed the probability to be as 4,000,000 to 1, that all the motions of the planets, both of rotation and revolution, were at once imparted by an original common cause, of which *we know neither the nature nor the epoch.*"

LARDNER.—This author, in speaking of the planets and satellites, says: "They obey the laws of gravitation, but they do much more. They all move in ellipses; those ellipses differ but very little from being circles; their orbits increase in distance from the sun nearly in regular progression; those orbits are nearly in the same plane; and their movements are in the same direction. Accordance so wondrous, and order so admirable, could not be fortuitous, and, *not being enjoined by the conditions of the law of gravitation*, must either be ascribed to the immediate dictates of the Omnipotent Architect of the universe above all laws, or to some general laws superinduced upon gravitation, which escaped the sagacity of the discoverer of that principle." Here, again, we see a reference to some unknown laws superinduced on gravitation, and another very strong denial that gravity can do what it really has done. The next author, like several others, speaks in the same strain.

NICHOL.—He declares that "not one of these remarkable arrangements in the solar system owes its origin to gravity. For instance, *gravity cannot account for the fact* that all the various orbs, primary and secondary, move in ellipses approaching very nearly to the circular form; nor the fact that all these orbs revolve in the same direction around the sun; nor the fact that they all rotate on their axes in the same direction; nor that equally singular ordinance which has confined so many bodies within a brief distance of the plane of the sun's equator. It appears a necessary conclusion, that the cause of the foregoing arrangements is *something profounder even than Newton's principle;* perhaps some remotest fact in the history of the universe."

HELMHOLTZ.—This author, in his very able essay on the conservation of force, and the application of that principle to the nebular theory, speaks of the motion of rotation as originally slow, and *the " existence of which must be assumed."* His profound views of the operation of the physical forces did not enable him to account for the origin of rotation; and he says, "the nebular hypothesis, from the nature of the case, must ever remain a hypothesis."

ALEXANDER.—In treating of a rotating nebulous mass this writer, with special formality, and therefore with due deliberation, says that "by a contraction from loss of heat there would result an increase of angular velocity of rotation, the *momentum* remaining unchanged." This is directly opposed to gravity as a force causing rotation, or even increasing the velocity; for if gravity act at all, the "momentum" of the nebulous globe must every instant be augmented. The origin of the rotation, or "the origin of the projectile force" in the mass that formed the solar system, he refers to a great primitive nebula which, by centrifugal force, threw off the portion which resulted in our system; but the origin of the rotation in that primitive nebula he does not conjecture.

"VESTIGES OF CREATION."—The anonymous author of this very wide-spread volume says, "It is a well-known law of physics that when fluid matter collects toward or meets in a centre, it establishes a rotatory motion," as in "the funnel, the whirlpool, and the whirlwind." This is simply the statement of a fact, not an explanation of the fact, nor the mention of the force causing the rotation of the stars. He says, gravitation and the loss of heat, two coöperating things, would cause the nebula to contract, and so do all other investigators, even when they deny in the strongest terms that gravity was the cause of the nebular rotation, and of the great leading peculiarities of the solar system. He also quotes and endorses Professor Nichol, where he says "the germs of the rotatory movement" are "obscure" and "vague." In fact, to understand the causes of contraction, is one thing; but to understand the cause of rotation and the increase in velocity of rotation, is quite another. There might be contraction without rotation, in the absence of the atomic repulsion force which held matter in the nebulous condition. Moreover, rotation could not possibly be originated at the centre of a nebulous globe. Rotation, as I have shown, must begin on the surface.

KIRKWOOD.—This author makes the following formal and deliberate statement: "In taking the most cursory view of the solar system, we cannot fail to notice the following interesting facts:

1. The sun rotates from west to east.
2. The planets move nearly in the plane of the sun's equator.
3. The orbital motions of planets and satellites are from west to east.
4. The rotatory motions are in the same direction.
5. The rings of Saturn move in the same direction.
6. The planetary orbits are nearly circular.
7. The cometary orbits have different peculiarities, etc.

None of these facts are accounted for by the law of gravitation. The sun's attraction can have no influence whatever in determining either the direction of the planet's motion or the eccentricity of its orbit."

When Professor Kirkwood proposed his "Analogy" respecting the periods of the rotations of the planets, his theory was supported by Messrs. Walker and Gould, and their statements contain the following:

WALKER.—He maintains by algebraic formula that in a contracting nebulous orb "the *momentum* of rotation is constant." "If a planet in a primitive state existed in the form of a ring, revolving around the sun, the *momentum* of rotation must, by virtue of the principle of conservation of movement, have existed in some form in the ring."

This is the general doctrine of all writers on this subject, as may be proved by many quotations. It is directly opposed to gravity as an animating principle in rotation, for that principle must increase the momentum every hour, as long as the nebulous mass contracts.

GOULD.—This author is here quoted because he does not ascribe the primitive movements to gravity, but refers to unknown physical forces. He says: "When we are considering the evolution of order from chaos, we cannot pretend to a knowledge of all the physical forces which exerted an influence."

ARAGO.—I have been unable to find that part of Arago's works where he treats fully of the nebular theory, as it is said he has done. His name is here introduced to show that, like several others, from Sir William Herschel until now, he believes in the operation of unknown forces on nebulous matter. He says: "Every thing authorizes us to suppose that the nebulous particles are subject, in the vast regions of space, to forces of which we have no idea."

HEINRICHS.—The leading peculiarities of the solar system are declared by this author to be relations and laws of a "*still higher order than those deduced from gravity.*" "The laws of Kepler are grand, as well as Newton's theory in accounting for them; but the above laws of Laplace are certainly of a superior order; and the theory of *gravitation*, *in failing to give a shadow of reason* for these laws, proves itself not to be the whole truth; *we must go beyond this force.*" The rotatory or projectile force he calls the "tangential force, of which *gravitation knows nothing.*" He is disposed to ascribe the rotation in the nebula to atomic force, "the grouping of several repelling atoms around one attracting atom."

TROWBRIDGE.—This author, like the author of the "Vestiges of Creation," seeks to find the origin of rotation at the centre of the nebulous globe, and from a variety of forces—"·the attraction of gravitation, the radiation of heat, and whatever other active and modifying cause might operate on the nebulous body;" again, he says, "a motion of rotation would result from the cooling down of the nebulous mass, the attraction of gravitation, and other forces of nature which might operate." Of course, this is not clear, and he says, "we have been unable to tell just how a rotation commenced." The reason of this is, because the commencement of rotation must be sought, not at the centre, but on the surface of a contracting nebulous globe, and from gravity alone.

MITCHELL.—The following extracts are forcible, and convey the ordinary opinions of scientific men: "In the outset of this description" of the nebular theory, "we must clearly distinguish between those phenomena for which the law of universal gravitation is responsible, and those other phenomena of the constitution of the solar system in the explication of which this law has never been employed. The solar system once being organized as it now is, all its existent and daily phenomena are susceptible of explanation from the theory of gravity."

"Here, however, the domain of this law is bounded; or, at least, has hitherto been bounded. There remains a multitude of inquiries demanding answers, for which, however, *gravitation has not been deemed accountable.* For example, why do all the planets and satellites revolve in orbits so nearly circular? So far as gravitation is concerned, they might as well have revolved in paraboles or hyperboles. Why do all the planets circulate about the sun in the same direction? How comes it that the planes of the planetary orbits are nearly coincident? *Gravitation renders no reply.* Again, the planets all rotate in the same direction in which they revolve. The satellites follow the same analogies, and even the sun itself is in like manner found to rotate on his axis in the same general direction." In this manner this writer goes on at considerable length, enumerating other grand features of the solar system, and declaring that gravity cannot account for their origin. The origin of them all, however, as we have seen, is directly traceable to gravity.

PANTECOULANT.—Mitchell, in connection with the above remarks, quotes some half dozen pages from this author, describing the operations of the nebular theory. These extracts seem to present his ideas in full, but they convey not the least hint that gravity was the cause of the stellar motions. I have not seen his works.

From this review, it appears that no previous investigators have entertained the suspicion or the conjecture that gravity was the cause of the motion of the stars. Their opinions may be classed as follows:

1. Sir William Herschel stands alone in proposing the action of gravity on the surface of a nebulous mass as the cause of rotation. But he neither possessed nor imparted the proofs. Much more is necessary to produce rotation than the simple subsidence of "eccentric matter." This would only produce several currents in different directions. A single current, acting alone, is necessary. That current must move down an inclined plane. That inclined plane must be formed of two things—the tangential motion of the current, and the gradual contraction of the nebulous globe. Moreover, he did not speculate on the nebular theory, according to which, the motions of the stars in their orbits are derived from rotation, be the cause of that rotation what it may. Nor did he ever, that I am aware, make an allusion to the formation of planets according to Laplace's theory, although his paper of 1811 was written at least seven years after the publication of that theory, and

how much longer I do not know. Neither did Laplace refer to the very different nebular theory of Herschel, which began to be published some twenty years before his own. Since then both these theories have been combined; though that which appears to me as the best part of Herschel's, has been unnoticed until now.

2. A few inquirers, such as Nichol, Trowbridge, and the author of the " Vestiges of Creation," sought for the cause of rotation in the meeting of the particles at the centre of the nebulous mass. This meeting of particles, and a rotation growing out of it somehow, they ascribed to " contraction from the radiation of heat," to " gravity," to " physical laws," to " other active and modifying causes," to " other forces of Nature," and to " unknown forces." No wonder that one of these authors quotes another in saying that " the germs of rotation " are " obscure " and " vague," and that still another says, " we have been unable to tell just how a rotation commenced." Even if any one of these authors had entertained the conjecture that gravity was one of the causes of rotation, it does not follow that he added the further conjecture that gravity was one of the causes of the projectile motions of the stars. The one idea leads toward the other, but the two are not necessarily in the mind together. I had the distinct idea that gravity was the cause of the rotation of a nebulous mass several years before the very natural consequence occurred to me that I had discovered the cause of the proper motions of the stars in their orbits. New ideas of great importance come out from the unknown darkness very slowly and with difficulty; but when they are once fairly out, the wonder always is, why they were not seen before. Probably I did not look for the cause of stellar motion. Like others, I was trying to learn the origin of the stars, and their peculiar arrangements in the solar system, from a condensation of vapor,

according to the nebular theory. The origin of their motions is another thing. The awful fact that our globe is flying through space at the rate of more than a thousand miles in a minute, seems repelling, and ascribable to God alone by His immediate miraculous power.

Even after rotation was conjectured by the above authors as arising out of a confused mixture of forces, known and unknown, then before the planets were seen distinct and complete, and flying with almost lightning speed around the sun, several other intricate processes and confused mixtures of forces had to be unravelled. There was the force of inertia, the centripetal and the centrifugal forces; there was the separation of many rings, the breaking up of these rings, their subsidence into round globes, the rotations of all those globes in the same direction, the formation of other rings for the production of satellites, the accounting for many other phenomena, and the answering of many annoying objections from keen and earnest opposers of the nebular theory. In all this, while the mind is puzzled to see the way clear through so many difficult operations, we may easily lose sight of an apparently small idea in the remote premises that gravity was one of the forces which concurred in causing condensation and rotation in some "obscure" and "vague" manner. Thus it is that an author may mention gravity as one of his forces to begin with, and lose sight of it in the protracted and multifarious process, before he gets his entire solar system in most rapid motion, where gravity has long been seen as the centripetal force, and the very opposite of the centrifugal force, the origin of which is now first seen also to be gravity. If the question then be put, What was the cause of the motion of the stars? he would have a hard task to retrace his steps and get back to gravity, mixed up among all his other beginning assumptions. Therefore, as no author has announced the distinct conjecture that gravity was the cause of the motions of the stars, I think we are not warranted in inferring that any such distinct conjecture was in their minds, merely because gravity entered vaguely, among many other things, into their speculations on the origin of the solar system.

3. By far the larger number of investigators have followed Laplace in assuming the rotation of the nebulæ as an unaccountable fact. They do not inquire into the origin of stellar motion any more than into the origin of matter. It is remarkable how many declare, in the strongest possible terms, that gravity could not have given motion to the stars. The extracts just cited on this point, especially from Newton, Nichol, Lardner, Kirkwood, Heinrichs, and Mitchell, are worthy of a reperusal.

SECTION XLII.

MODERN THEORIES OF CREATION.

THE creation of the universe of stars, and especially of the star on which we dwell, is a solemn wonder to the human mind. It has always been a matter of the deepest and the most serious inquiry, and so it will remain forever. Assuming that the theories of the origin of the stars, of their motions, and of their light and heat, are true as detailed in this volume, we may take a brief review of these and of other true theories of creation, and thus arrive at the grand epic idea which their union affords. The theory derived from the ancient philosophers, that our earth is round, in the form of a globe, must be regarded as the grand starting-point of all modern theories. It was an immense advance beyond the ideas of the more ancient Chaldeans, Egyptians, and Hebrews, who regarded the earth as a level plain, indented by a few seas, and surrounded by a vast unknown ocean of waters.

1. The first great theory of modern times was that of Copernicus. Our world, instead of being the centre of the universe, around which all the celestial bodies revolve, becomes by this theory a very small star, revolving with double motion around another great star, and both these motions with most strange and wondrous velocities. Probably no truth ever presented to the human mind appeared so incredible.

2. A new theory was necessary to establish that of Copernicus. He believed that the moon revolved around the earth, and the planets around the sun; but these revolutions must cause a vast amount of centrifugal force! What prevented the moon and what prevented the planets from flying away out of their orbits? A FORCE was needed to hold the solar system together. At first this serious difficulty was overcome by the reverent belief that the same Almighty Being, who gave the stars their motions, also held them in their orbits. But God everywhere works by means, and the theory was formed that they are held in their orbits by the familiar force of gravity. When and by whom this idea first arose I do not know. Newton names at least six persons by whom it was entertained before his own publications, and some of these had published the theory to the world. But Newton discovered and proved to be true what was before a mere surmise. His chief merit lies not in proving that the centripetal force of gravity equals the centrifugal force, and that therefore gravity is the force which retains the stars in their orbits. His great superiority is seen in his applications of the theory of gravity to account for various astronomical phenomena, such as Kepler's laws, the precession of the equinoxes, and very many others. This made his "Principia" the greatest of scientific books.

MODERN THEORIES OF CREATION. 345

His mathematical reasoning was most profound, and far in advance of what the human mind had before achieved. He also added to the theory of gravity the important element that, instead of acting with equal force at all distances, that force decreases inversely as the square of the distances.

3. The next great theory of creation is the nebular theory. Undoubtedly the credit of its formation belongs to Laplace. By profound meditations he saw that many features of the solar system must owe their origin to a common cause. The rotations and revolutions are all from west to east, all nearly in the same plane, and that plane not far from the plane of the sun's equator. The rings of Saturn have the same movement, in his own equatorial plane. He saw the probabilities to be exceedingly strong, so strong as to amount to a certainty, that all these phenomena must have had the same origin; and with wonderful sagacity he referred that origin to a former nebulous envelope of the sun which had extended beyond the planetary orbits. The temporary star of 1572 suggested an extraordinary amount of heat to produce this nebulosity. Even the absence of comets of short periods, now known to be an error, was proof to him; for, from the analogy of the planets, he inferred their former existence, and that they had been destroyed by the sun's temporary nebulosity. By a gradual cooling, he thought that rings had been parted from the equatorial zone of the sun's nebulous atmosphere, that these rings had settled into nebulous planets, which in cooling and condensing had, in like manner, parted their satellites. He did not attempt to account for the cause of the sun's rotation, nor for the origin of the sun, or of the fixed stars, or of the comets or meteorites. All that he wrote on the nebular theory amounts to less than a half dozen pages of an ordinary book, and it is all printed in Appendix V. to this

15*

volume. Small in bulk as it may appear, it is yet most valuable and meritorious, for it has been the precious germ of what the nebular theory has now become.

4. But a new theory was necessary to establish the nebular theory of Laplace. Just as the theory of Copernicus was deficient for want of a force to hold the solar system together, so the nebular theory was deficient for want of a force to give motion to the solar system. Even assuming as an unaccountable fact the ·rotation of the sun, still that velocity of rotation, in the cooling and contraction of the sun's atmosphere, could never give sufficient centrifugal force to balance the centripetal force, and therefore no ring at the equatorial zone could be parted. Even if the rotation were rapid enough to part a ring at the orbit of the outermost planet, still not another ring could ever be abandoned. The reason is, that at the orbit of the next interior planet, the centripetal force is much stronger, and therefore the centrifugal force must be just so much stronger. But this stronger centrifugal force could not be obtained without a more rapid movement of the equatorial zone of the sun's atmosphere. This movement must be as much more rapid at every planetary orbit as one planet flies more swiftly than another. While at Neptune it must have moved 12,500 miles an hour, at Mercury it must have moved 110,000 miles an hour. It has all along been seen that the angular velocity would be increased by the contraction of the nebulous mass, but this would not touch the difficulty. An increased linear as well as an increased angular velocity is absolutely necessary. No force was known to impart this increased rapidity of motion. Neither could Laplace nor any of his followers understand how the sun's atmosphere could have a velocity of 12,500 miles an hour when at the orbit of Neptune, 110,000 at the orbit of Mercury, and only 4,500 at the present era. Nor was any method

known for calculating what, in the process of contraction, the velocity of rotation should be.

The theory of the present volume supplies the force necessary to establish the nebular theory. By regarding gravity not only as the original cause of rotation, but as the cause of the increased velocity of rotation, we can calculate, from time to time, how rapidly the nebulous mass must have rotated, and then we discover that the rotation must have coincided with the present planetary velocities, after all necessary friction is allowed. That same friction accounts for the decreased velocity of the sun's equator in the present era.

Thus, as gravity supplies the force necessary for the Copernican theory, so it supplies the force necessary for the nebular theory. In the one theory, gravity gave the stars their motions in their orbits; in the other, gravity retains them in their orbits. As the investigations by Newton in the action of gravity proved the Copernican theory to be true, so I trust the investigations in the actions of gravity, as presented in this volume, prove the nebular theory to be true. Both theories are alike impossible without the foundation of gravity.

5. By light all creation becomes known, and by light and heat all life, both vegetable and animal, is supported. Therefore, the origin of the light and heat of the sun and stars, enters our very first ideas of creation. No wonder that, in the most ancient and venerable of all books, the creation of light takes precedence of all other things. The theory that chemical action, or ordinary burning, causes all stellar light and heat, dates from time immemorial. It was held by such great men as Newton and Laplace, and I suppose also by Davy, for to chemical action he ascribed the former fiery condition of the star on which we dwell. Lately it has been totally abandoned by all scientific men, but only

on account of a single difficulty. The objection is precisely on a level with those which have been made against the Copernican and the nebular theories; namely, the want of a FORCE to keep the process of the theory in operation. Where, say the objectors, is the force necessary to hold the solar system together? Where the force to put it in motion? And where the force to keep up the chemical action in the sun so long? If its materials are like those of our ordinary fires, its burning must soon cease, for want of fuel; but geology reveals that it has been sending forth light millions of years. Nevertheless, the proofs of the chemical origin of stellar light and heat are overwhelmingly strong, and the true scientific course, therefore, is not to deny the chemical action of the stars, but to search for the origin of the chemical force.

6. The nebular theory and the recent theory of the conversion of forces afford the foundation for a new theory of force, by which the chemical action of the sun has been so greatly prolonged. According to the nebular theory, as presented in this volume, all the space within the ring of the Milky Way was occupied by matter in a nebulous condition, which broke up into stars. The matter of our own solar system, nearly all of which is now in the sun, occupied the space until half way to the nearest fixed stars. Assuming those to be a little more than 20,000,000,000 miles distant, then the space occupied by the sun extended more than 3,622 times farther every way than the distance of Neptune. But we have already seen that the sun, when expanded to the orbit of Neptune, was 14,000,000 times less dense than hydrogen. Therefore when expanded 3,622 times farther, it must have been 666,000,000,000,000,000 times less dense than hydrogen. This great number expresses the repulsive force which then resided in the sun and separated its particles. What has become of this

force? It has not been destroyed, for force, like matter, is indestructible. But it may have been converted into some other force, the same as the centripetal force may be converted into the centrifugal force, antagonistic to itself. We may suppose that all this inconceivable amount of repulsive force has been converted into chemical force, also antagonistic to itself. We can think of no other way for its disappearance. We need not inquire how this conversion from repulsive to chemical force could take place, for we know not how any conversion of force takes place, except the conversion from the centripetal to the centrifugal force. We can thus understand how there may be an indescribable amount of chemical force deposited in the sun. It must have been there, if the nebular theory and the theory of the conversion of forces be true; and of their truth a doubt can no longer survive. A large part of this chemical force residing in the sun has been expended or converted into light and heat, probably by far the larger part; but how much still remains, we have not the least grounds for forming an opinion. This large reservoir of chemical force in the sun may operate in producing light and heat by three different ways.

a. The force may be latent in chemical elements already formed. In oxygen and hydrogen there is a latent chemical force, which appears when these two elements combine to form water. In some elements a larger amount of this chemical force is latent than in others. How much force is latent in the peculiar elements of the sun, we have absolutely not the least idea—perhaps enough to afford our present supply of heat and light for many millions of years.

b. The second way by which the vast reservoir of force in the sun may operate in producing an extraordinary supply of heat and light, consists in the *condition* of aggregation and an intense activity on the part of all the ethereal media.

It is remarkable how many forces or ethereal media are always at work in powerful chemical action. Electricity, magnetism, heat, and light, are all active; and even gravity enters in among the rest to perform its part by simple pressure. When all the known physical forces, all these ethereal media, are at work together in so vast a body as the sun, 882,000 miles in diameter, there must be an aggregative and reciprocal conflict of force, there must be an intensity of operation, which we cannot begin to comprehend. What may be their combined effect in producing heat and light, we have not the slightest means of knowing from our small experiments. The effect of even a single force on a large scale cannot be inferred from its effect on a small scale. We could never have known that a large magnet can cause gold and silver to point diamagnetically, from seeing a small magnet. Neither could we have known that potash, soda, and lime can be decomposed into oxygen and the metals, from experimenting with a small galvanic battery. The great battery, or rather the great laboratory of the sun, most undoubtedly produces effects in the form of light and heat which go beyond all our anticipations. For us to pretend to measure the heat-producing power of the combined action of all the physical forces in chemical action there, by our small operations here, seems most absurd. I speak with all due respect for the eminent men who appear not to have thought of this.

c. The third way by which the great store of force in the sun and stars, may operate in producing heat and light, consists in the formation of new combustible elements. New chemical elements have been forming all along during the nebular condensations. Of this we have powerful evidences in the eight classes of facts already described. In what ways these new elements are produced, we know not; but we can conceive of two. They may be formed of other more simple elements, the same as we form compounds. They might then be simple substances to our means of analysis, but they would be in reality compounds. This view becomes probable when we learn from the nebular theory that matter has been condensing by chemical action from a condition 666,000,000,000,000,000 times more rare than hydrogen. It must have combined over and over again many thousands or millions of times; and what we call, simple elements must be almost infinitely complex. Again, new combustible elements may, perhaps, be created out of incombustible compounds by means of the physical forces, the same as elements may be changed into allotropic states by the physical forces. In this case, however, as much force would, perhaps, be expended in making them combustible as could be derived from them in the form of heat and light. But the amount of force is not the difficulty. That we know exists in the sun, and we daily feel its power. The precise thing which we want is not force, but force in operation as chemical action.

Thus the Copernican theory, the nebular theory, and the chemical theory of stellar light and heat, were all during long periods discredited for want of forces to keep them in operation. The inconceivable amount of repulsive force which held matter in a diffused nebulous condition, and which we may now regard as converted into chemical force, supplies what has been wanting in the chemical theory, and gives it a firm foundation.

7. The next great theory of creation was to account for the structure and formation of the solid globe on which we live, including the origin of the continents, the islands, the mountains, the varieties of rocks, and the fossil remains of plants and animals. After centuries of observation and deep thought by many men in all countries, the conclusion was reached unanimously in the beginning of the present

century, that the interior of our earth is in a highly heated condition, that the solid crust is composed of either igneous or aqueous rocks, that the igneous rocks have cooled and hardened from a former melted state, that the aqueous rocks have been deposited as sediment in the bottom of waters, that the continents and islands have been raised up by mighty forces above the waves, and that the fossil remains of plants and animals are the relicts of the former inhabitants of our planet, which flourished at very different eras during many millions of years in the past.

This was one of the grandest acquisitions ever made, but its great defect was in not accounting clearly and definitely for the forces which had done all these wonders. The forces themselves were pretty generally known, but about their action the most extravagant opinions were entertained, such as the instantaneous rising up of continents and mountains, the sweeping of great ocean-waves over the lands, the sudden extinction and again the equally sudden creation of tens of thousands of species of animals and plants.

8. The next great theory was about the FORCES which had produced all these vast geological phenomena. It was originated, as far as I am aware, and ably explained and advocated, by Sir Charles Lyell. In his "Principles of Geology," he showed, to the satisfaction of all geologists now living, that all geological phenomena had been brought about very slowly, and without any sudden general catastrophes, by the ordinary forces which are now around us in daily operation. It has been truly said that " the publication of ' The Principles of Geology ' forms one of the eras in science."

Some readers may think at first view that this is giving too much credit to Lyell. All the forces, the igneous, the aqueous, the chemical, the atmospheric, the solar in heat, the lunar in tides, the organic in animals and plants, had been discovered long before. But these forces were supposed to have acted with far greater intensity, on a much grander scale, and with an unknown irregularity in former geological eras. Hence former writers spoke of universal catastrophes, the sudden sinking and the sudden emergence of continents, the instantaneous destruction of all animal and vegetable existence, and again their sudden and instantaneous creation. This wild confusion, this chaotic uproar, this sweeping of great waves over continents, were no longer mentioned after the appearance of Lyell's " Principles of Geology." By this we see how great an advance was made by the discovery that the ordinary forces now in operation, in their present intensity, in their present manner, are sufficient to explain the origin of all the facts in geology. The islands and the continents have doubtless been raised above the waters, and their mountain-peaks have been reared far above the clouds, by the very forces which are now raising them up little by little. We can trace the sea-beaches one after another to higher and higher levels, and other evidences bring us at length to the loftiest mountain-tops as the results of a slow and nearly continuous process of elevation.

9. Another grand theory accounts for the creation of the many species of plants and animals, not only of those now living, but the far greater number revealed by geology in former epochs of the world's history. All believe that the creation of all things, in every department of the universe, is the work of a Wise and Benevolent Being; and that these exquisite organizations of plants and animals have been designed and brought into being by His direct superintendence. But the great question arises, How has He done this? Was it done from time to time, in the successive periods of the world, by sudden exertions of miracu-

lous power, or did He work by means and according to laws? There are the most powerful and convincing reasons for believing that the various species of animals and plants were originally formed by law and by means, the same as individuals of all species are now created by law and by means. This, to use an old Hebrew expression, is "the way of the Lord." So He created the stars, so He created the geological structure of our globe, and so we may naturally suppose He created organic beings. At the present era there are about 500,000 species of animals and plants alive on our globe. Of these, about 150,000 are plants, and about 300,000 are insects. I suppose our world has been gradually emptied of one set of its inhabitants, and as gradually and insensibly filled with another set many thousands of times; the outgoing of the one and the incoming of the other proceeding simultaneously. But on the supposition that there has been only a hundred different sets of inhabitants on our globe, a supposition which no geologist will deny, then there must have been created in all 50,000,000 of different species. Now it is our uniform experience, that when God creates things by the tens of millions, or by the hundreds of millions, He does this not by miracles, but by general laws and by specially adapted means. When, for instance, He creates, every season, for the sustenance of man, many millions of grains of wheat, He performs no miracle; He works by means and according to laws. This is His "way." We can think of no possible reason why He should have departed from His way, to perform 50,000,000 of miracles in creating plants and animals. If God performs a miracle, we may be certain that He has some well-known object in view, but we can think of no object for the performance of these 50,000,000.

Assuming, then, that all species have been created by

specially adapted means and according to general laws, we can think of only one process of their creation, and this is, that one species was produced by means of another almost precisely the same. This is called the theory of the transmutation of species. There are about three hundred species of doves, about three hundred species of humming-birds, and in the United States alone, about six hundred species of fresh-water mussels; and in the same country there are an equal number of species of water snails called melania, or nearly allied to melania. There are in all about two thousand species of land snails, and about five thousand species of mushrooms. There are many genera of plants whose species are numbered by the hundreds. In all these genera, the several species so nearly resemble one another, that ordinary observers cannot tell the differences between them. In this way the theory of the transmutation of species supposes there have been insensible gradations between all living things, and that they all may be traced back little by little to simpler and simpler states, until they arrive at last to the same form, a round cell. Geology reveals the great fact that the first created plants and animals were very low in the scale of organization, and that in every new epoch in the world's history, higher and higher forms have appeared. We are therefore led to believe that the first organized living things were simply round cells in the water. By what means God created these, and endowed them with what is called vital force, no one knows as yet. These round cells varied so as to become plant cells on the one hand and animal cells on the other. The animal cells then varied into four different departments, the radiates, the molluscs, the articulates, and the vertebrates. These departments each varied again into three or four classes. The classes next varied into different orders, and the orders into genera, and the genera into species. In this very simple and natural manner we can account for every peculiarity in all plants and animals.

Who originated this theory I do not know. It has been advocated in the present and in the preceding century, but its history through the distant past has not been written. Until quite recently the weak point of the theory was, that it could assign no satisfactory FORCE for the changes from one species to another. Some appealed to outward physical conditions, such as land and water, cold and heat, dampness and dryness, hills and plains, forests and meadows, fruitful lands and deserts, rocks, mountains, valleys, and the like. Others thought that the strong inward feelings and desires of the individuals caused the outward shape as well as the internal organization to change. One author thinks that God so contrived all species, that like should produce like until such time as He had ordained a change; and that then, by a small variation, a new species should be produced, which should continue permanent for another long period, until its own turn arrived for producing another change. But all these causes for the change from one species to another were unsatisfactory; and for want of an adequate *force* for carrying on its process, the theory remained neglected. The creation of new species from time to time in the world's history was regarded as the "mystery of mysteries."

10. The honor of discovering the force for changing one species into another belongs to the profound and devoted naturalist CHARLES DARWIN. He calls it variation and selection. It is a familiar fact that no offspring is precisely like its parent, and that no two animals of a litter or family and no two plants of the same species are alike. Variation within small limits is the universal rule. Any individuals that vary in a way that is advantageous, will be better able to prolong their existence, when their paternal

forms are left to fall a prey to their living enemies, to the violence of the seasons, and to the scarcity of food. Therefore, these new forms are said to be *selected* for living and propagating. These in their turn will inevitably produce other new variations, some of which will be still more favorable for prolonged existence; and these again, in the competition or battle for life, will be selected to live and propagate when their paternal forms are unable to maintain their existence. Thus, in adding little by little hundreds of times over to any one species, any amount of changes may occur. These changes are not caused directly by any external conditions. They arise, first, from the universal tendency of the vital force for *variation;* and, secondly, from the *selection* of improved variations to live and prolong their race. This law of variation, and this inevitable selection for life of the most improved variations, must be regarded as the force for producing new species. When the species become very much changed, they are called new genera; and when the genera change, they become new orders.

While we can trace resemblances between all living things, even between plants and animals in their very lowest forms, there are nevertheless lost links in the chain of being, wide gaps here and there between animals most nearly related. Between the dog and the bear, and between the dog and the cat, there are large spaces; and any one of these forms could not possibly have produced another. This is admirably explained by theory. These lost links, these wide gaps or spaces between different forms, represent many forms that have been crowded out of existence. These extinct forms were far more numerous than the forms now living. Let us suppose, for illustration, that from the carboniferous period until now, there have been one hundred and one different sets of inhabitants on our globe. Then, as one hundred of these sets are dead, and only one living, the number of perished forms must be to the number of living forms as one hundred to one. Therefore the lost links in the chain of being, those wide gaps between different living forms, are evidences which favor the theory of the transmutation of species. They are inevitable consequences of that theory, and confirmations of its truth. The fossil remains confirm this view. They are being discovered by hundreds every year, and they are constantly restoring the lost links in the chain of being, constantly filling up the gaps between one form and another. Quite recently a petrified bird was found, with a long tail of several vertebræ, like a dog or lizard, each vertebra having a pair of large tail-feathers. It is extremely improbable that all the lost species will ever be discovered in a petrified state; but the evidence of geological discoveries, so far as it goes, is perfect in favor of the transmutation theory, because the discoveries, as fast as they proceed, are constantly restoring the genealogical tree by filling up the vacant spaces between kindred forms.

Here, for want of space, must cease our review of modern theories of creation. They are ten number, and it is plainly seen that they divide themselves into two classes: those of operation, and those of force to carry on the operation. The five theories of operation are the Copernican theory, the nebular theory, the chemical theory of stellar light and heat, the geological theory, and the theory of the transmutation of species. The other five are theories of force, because they are employed about the powers by which the known universe was created, and by which it is now upheld and moved in admirable order. In all these cases the theories of operation were formed before the theories of their respective forces. This is the natural order of discovery as it proceeds from effects to causes. But in every case, when any theory of force has been formed, it has added instantly and largely new light to its corresponding theory of operation. Newton's theory of force added wonders to the Copernican theory; Lyell's theory of force, immediately and in his own hands, gave an onward movement to the geological theory; Darwin's theory of force, at once and by his own application, established on an enduring basis the theory of the transmutation of species. It is hoped that the formation of two new theories of force, as explained in this volume, will add their due contributions, both to the nebular theory and to the chemical theory of stellar light and heat.

From all these theories it appears that the government of the world—the cosmos—is now carried on by the same forces by which the world was created. Indeed, the creation and the government of the world is but a single continuous act. The same gravity which put the solar system in motion, still controls that system. The same chemical force which lit up the stars, still makes them shine. The same geological agencies which fashioned our globe, are still laying down new strata, raising them up, and moulding their surfaces. We know not, indeed, the forces employed in the formation of the first organic cells, but we know that the same forces by which all plants and animals have been evolved from simple cells, now animate and control them.

SECTION XLIII.

THE HISTORY OF CREATION.

In writing the history of the past, we can go back to the time when all matter was diffused through all known space. There it hangs as a veil, hiding from our view all the previous doings of the Eternal. There let our farthest point of vision rest contented until clear and well-founded revelations enable us to look beyond.

In this diffused nebulous matter contraction began, and it broke up into immense masses. Each one of these resulted in a stellar system. They now appear as faint patches of light on the dark ground of space, and good glasses resolve a few of them into millions of stars. Some of these patches are called globular, some elliptical, some planetary, and some annular nebulæ; and others, with a great star in the centre, are denominated nebulous stars.

Our own stellar system consists of a ring of stars called the Milky Way, a disk of stars within, and scattered nebulæ without, that Way; both the disk and the scattered nebulæ lie in the plane of the ring. But disk, ring, and scattered outer nubulæ at first formed a single very irregular nebulous mass. It subsided into a globe, rotated, abandoned rings one after another, and spread out its small remnant into a central disk. The outer rings broke up into several large masses, and then each mass separated into a system of stars, forming the outer scattered nebulæ. The inner ring and the central disk preserved their general shapes, but each one also separated into smaller masses, which formed smaller clusters of stars or solar systems.

Our own solar system, when in the condition of a very rare nebulous mass, was not very irregular in shape. Soon it became round, rotated, and abandoned rings. First its rotation was extremely slow; then by degrees it increased its velocity to 12,500 per hour, when the ring forming Neptune was abandoned. Its speed was still increasing, for its surface was gliding down an inclined plane under the continued action of gravity; and at last, after parting many rings, it reached the enormous and almost incredible velocity of 110,000 miles per hour. Then the ring resulting in Mercury was parted; but no more rings were parted, because no greater velocity of rotation was reached. The interior of the mass had been settling in radial lines toward the centre, and without rotation. The time arrived when condensation had reached a point that necessitated a greater friction between this unrotating interior and the swiftly-rotating exterior. Hence the exterior was greatly retarded, and this retardation continued until now the equatorial velocity of the sun is only 4,564 miles per hour.

The ring from the nebulous sun, giving origin to our earth, broke, like all the other solar rings, and became a contracting, rotating, nebulous globe. This latter nebulous globe was large enough to abandon only a single ring, and that, after subsiding into a rotating globe, became our moon.

We have seen reasons for believing that the contraction of nebulous matter was caused by chemical combination. This combination produces heat and light, and hence the light and heat of the sun and fixed stars. But it is the nature of chemical action in any body to come to an end; and we see, therefore, that as the stars condense, the light of one after another goes out, and they become lost stars. All the members of our solar system, except the great central one, have lost their light, and they are now opaque. Chemical action on so grand a scale as an entire star must evolve heat enough to hold matter in a liquid condition; and hence the present liquid condition of the sun and fixed stars, and also the former fused condition of the earth and the moon.

But it is the nature of a heated body to radiate away its heat, and hence a solid crust began to form on the moon and on the earth. This crust was first in patches, floating about, as ice first congeals on a river surface. They were at a temperature of white heat, like the spots on the sun, which are still less brilliant than the surrounding liquid, and therefore seem dark by contrast. These solid spots floated, because all solid rocks, like lavas, float on the surface of the same rocks when melted. The same is true of cast iron. At length, as they multiplied in number and grew in size, they completely enveloped the liquid globe whose light was now extinguished, except here and there, faintly and temporarily, in volcanic eruptions. The moon being the smaller, and having no deep atmosphere heavily charged with watery vapor, cooled more rapidly by far than the earth; and hence the lunar volcanoes are so much more numerous, so much larger, and all long ago extinct.

When the solid crust of the earth was still red hot, all the waters were floating in the form of vapor in the atmosphere. With such a coating, how slowly, according to Professor Tyndall's researches, must the earth's heat have radiated away! But as it cooled, the vapor condensed, settled down, and became the water of the ocean, completely enveloping the solid globe. Whoever, near an iron furnace, has seen the solid, glassy slag, anhydrous, and overcharged with lime, crumble into a pasty mass at the first fall of rain, can imagine that much of the solid crust of the earth, anhydrous, and overcharged with alkalies, must have dissolved into a soft, muddy mass when the waters first came down. That mass became the ready material for being afterwards washed away by ocean currents, and laid out into stratified aqueous rocks.

The heat of our globe continued to radiate away, and contraction followed. But the liquid interior, like all liquids, contracted in volume more than the solid exterior crust. That crust, therefore, became too large. Hence it wrinkled into furrows, to suit the diminished size of the globe. The downward bends of the furrows formed the ocean floors, and the upward bends became the islands and continents. At first the lands were low, level, and moist; then the surface became more undulating as the furrowing increased; then hills and mountains were raised up; valleys were scooped out by running waters; and at length, after millions of years, its present surface was assumed.

Life began at a very early period in the waters when they were still warm, and covered the entire globe. Vegetables and animals, we suppose, were at first only simple round cells, and afterwards became more complicated and varied. They were all marine. The atmosphere was too heavily charged with carbonic acid to support air-breathing animals, except the very lowest amphibian kind. But in process of time this gas was absorbed by the lime and by the land plants, forming coal and carbonate of lime. This carbonate of lime, after being employed for the hard parts of the lower animals, now appears as chalk, limestone, and marble.

Thus was the air fitted for the higher orders of animals. The first air-breathing animals were of the lowest air-breathing forms, and allied to the frog, which, in its early days, has the character of a fish. Afterwards higher and higher structures of reptiles were produced. Then came the birds, though at first they had long tails like reptiles. Then higher and higher orders of birds, and last came the mammals, the highest class of all. But the lowest orders of this class first appeared, such as the marsupials and the pachyderms. Then came the ruminants, then the carnivorous orders, and then the fruit-eating monkeys, sporting in the tree-tops. Man appeared last on the globe. His antiquity dates back, we cannot say how long, but it is certain that his years must be reckoned by the hundred thousand.*

By all this history it is evident that the nebular theory and the geological theory contemplate but a single process. The point cannot be defined where the one ends and the

* See Lyell's "Antiquity of Man," and recent works by other authors.

other begins. The ocean was nebulous long after the land was solid. We can trace a connected chain of events, beginning at our own era, and going back through the long ages that are past, until we arrive at the period when all matter was diffused everywhere in a nebulous condition, many thousand billion times more rare than hydrogen.

Such is the grand process of creation. It is not yet done. It is going on rapidly in our own day—more rapidly than ever. Changes in the earth's structure and in its inhabitants are constant. New strata are being laid in the waters, new deltas are forming, new deposits are made on subsiding regions, and other regions are bodily rising up and suffering denudation. The agency of civilized man is an additional force for geological changes. Five hundred millions of tons of coals, and ores, and rocks, and clays, and other minerals, are annually raised from the earth, and transported and changed. The plough, the spade, the hoe, and the harrow are only beginning their mighty work. And this work will be increased many fold when driven by steam and electro-magnetism. These agricultural implements mellow the earth's surface and clear away a matting of forest and other minor roots; thus the rains are assisted to carry off greater quantities of sediment, to level the prominences, to fill up the depressions, and to convey the earthy materials far away to the ocean-beds.

Corresponding changes are necessitated in the vegetables and animals. Every intelligent botanist knows how many tribes of plants are migrating to regions where they were before unknown. With the plants move the animals that live on the plants. Man, " the mighty hunter before the Lord," is causing many tribes of animals to disappear. Those which remain acquire new instincts and new characteristics to adapt them to the change. With civilized man come new animals, either subjected to his use, or parasitic upon him, like the rat, or like many species of insects that live on his cultivated plants. Man himself is changing. Intellectual, refined, and living more by his ingenuity than by his strength, he is a different being from the savage, in whom passion and physical force are the chief traits. This change in man is also now going on more rapidly than ever. Reason, humanity, and intellectual exertion are becoming more and more prominent. In the use of the physical forces, we are just learning what it is " to have dominion over the earth and to subdue it." Steam, electricity, the printing-press, the paper-mill, and a thousand other new arts are changing human conditions, human employments, human habits, and human characteristics. Mind is becoming more and more the standard of the man.

As when a ball rolls down an inclined plane, and acquires constantly new velocity, so the sublime process of creation appears to be a constant rolling down, a daily expenditure of force, and it now progresses with a greater velocity than in former eras. The human mind is becoming one of the great forces of creation. With the Wise One of Galilee we can say, " My Father worketh hitherto and I work." With Paul we can exclaim, " Now then we are workers together with God!"

SECTION XLIV.

THE CREATION AND GOVERNMENT OF THE WORLD BY SPECIAL FORCES.—DIVINE PROVIDENCE.

From the facts and reasonings of this volume, it appears that the creation of the world, the forms and the arrangements of the material universe, have been brought about by special forces. These forces are still in operation, and by these the world is now moved and governed.

In all things we at once perceive the difference between

design and chance. How quickly do we recognize chance in a thousand bricks thrown carelessly on a heap; and how quickly do we recognize design in the same bricks carefully built in a wall! We need but open our eyes at any hour, and the multitudes of objects around us divide themselves into the one or the other of these two classes. Wherever we see an object adapted for certain ends or purposes, there we see design. Through all creation the instances of design are innumerable. How wonderfully are the human arm and hand adapted for special purposes! How wonderful a piece of machinery is a tree for producing flowers and fruit! How wonderfully is the sun adapted for the well-being of all plants and animals which rejoice in his rays! Where there is contrivance and design, there must be a contriver and designer.

All creation everywhere is the work of design; and as creation has been brought about by the agency of special forces, therefore these forces are the work of design. They have been contrived or designed for producing special purposes, the beautiful and sublime forms of the universe.

All creation is now governed by the same forces by which it was produced. The government of the world and the creation of the world is but a single piece of history, a single act. And as its creation was by design, therefore its present government is by design. The same Contriver and Designer who presided over its origination, now presides over its movements. This Originator and Upholder, this Contriver and Designer, we call God, the First and the Last, the Beginning and the Ending; who is and who was, and who is to be, the Almighty.

APPENDIX I.

A MATHEMATICAL PROCESS—TO XVIIITH SECTION.

THE process for finding the velocity of rotation of a contracting nebulous sun, at different periods in the history of its contraction, consists of two parts, as follows:

First Part.—Find the velocity of a fall from infinite space to the orbits of the several planets, regarding the sun from time to time as expanded to those orbits. That velocity would be $= \sqrt{2gr}$; in which formula r stands for the sun's radius, or, in other words, the planet's mean distance; g stands for the velocity, in feet per second, of a fall at the end of one second on the sun's surface, when so enlarged. This latter velocity is found from that at the earth's surface, which is 32.16 feet per second, by comparing the mass and radius of the expanded sun with the mass and radius of the earth.

Second Part.—Let M be the velocity of a fall from infinite space at the orbit of any planet, and let N be the velocity of a fall at the orbit of the next interior planet. Then N — M will be the velocity gained by a fall from the former to the latter orbit. Then N — M added to the *actual* velocity at the former orbit, will give the velocity at the latter orbit as due to gravity. From this latter velocity subtract the actual velocity of the planet at the interior of the two orbits, and the remainder will be the retardation by friction on the inclined plane between the two orbits.

The reason for this Second Part of the calculation arises from the retardation by friction. This friction is an absolute necessity in the nebular theory. It is an essential part of that theory, and without which the theory cannot stand. Therefore, any calculation or demonstration on planetary velocities, ignoring this friction, or this Second Part in the above process, is worthless. What would be said of any mechanical engineer who ignored friction? The solar system is a mechanical production, and in its formative process, when its parts were moving on one another, friction was clearly inevitable.

APPENDIX II.

ANSWER TO A MATHEMATICAL OBJECTION—TO XIXTH SECTION.

THE following demonstration was presented by some three or four very intelligent mathematical gentlemen as an objection to the conclusions in this volume on the cause which gave motion to the stars. It is placed here, with a rejoinder, to prevent the same objection from arising again in any other way hereafter. It may occur independently to other minds. The error of the objection arose from ignoring the Second Part of the process explained in Appendix I.

DEMONSTRATION.

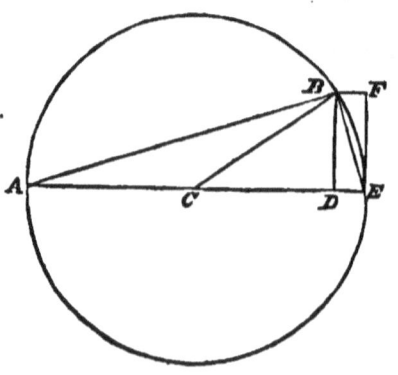

Let ABC be the orbit of a planet. EB an arc passed through by a planet in 1 second.
Then $ED = BF =$ distance fallen toward the sun in 1 second $= \frac{1}{2}g$.
$AE =$ twice distance of a planet $= 2r$.
BE (very nearly arc BE) velocity of planet in orbit $= v$.
$EA . BE :: BE : DE$.
$2r : v :: v : \frac{1}{2}g \therefore v = \sqrt{gr}.$
But velocity acquired by fall from infinite distance $= \sqrt{2gr}.$
$\therefore$ Planet's velocity in orbit : velocity from fall from infinite distance $:: 1 : \sqrt{2}.$

APPENDIX II. 369

This relation being essential to stability, must exist, *whatever* be the origin of the velocity. Hence it proves nothing as to the source of the orbital velocity, except that it is entirely *compatible* with the assumption that it is due to gravity.

REJOINDER.

This relation between any orbital velocity and a fall from infinite space to such orbit, cannot be said to be " essential to stability." Nothing more nor less is essential to stability in the case here mentioned than an orbital velocity which shall produce a centrifugal force equal to the centripetal force. The even balance between these two forces is *the fact* essential to stability. The *constant relation* between the orbital velocity and the velocity of a fall from infinite space " must exist " indeed, but this is an additional and another different fact, and exterior to the precise thing which is " essential to stability."

This constant relation is truly " compatible with the assumption that the orbital velocity is due to gravity," but so far from " proving nothing as to the origin of that velocity," it proves a very great deal. It proves that now we are acquainted with a physical force which, acting in accordance with the nebular theory, is powerful enough to produce velocities as great as any of the orbital velocities, and also in such immense bodies as the planets, and again in the very directions in which the planets move. What makes this proof more important is, that we are acquainted with no other physical force capable of doing any of these things. Moreover, that force would not be powerful enough to produce any one of these orbital velocities very far beyond the orbit; and if it be too powerful, abstractly viewed, to produce any one of these orbital velocities in the region where it moves, yet if viewed as acting in accord-

16*

ance with the nebular theory, then that force and those several velocities perfectly coincide.

The relation between the orbital velocity of a planet and the velocity of a fall from infinite space to the orbit of that planet, viewed abstractly and alone, and independent of friction, is not taken in this volume to prove the origin of that orbital velocity. There is a discrepancy between the two as 1 to $\sqrt{2}$, or as 1 to 1.41. But in ascertaining whether gravity be the origin of the orbital velocity of any planet, we must proceed in accordance with the nebular theory; we must compare the actual velocity of the planet, not with the velocity of a fall from infinite space independently of any other consideration, but with the velocity of a fall from the orbit of the next outer planet added to the actual velocity of that outer planet. Take the earth, for example. Its actual velocity is to the velocity produced by gravity in a fall from the orbit of Mars added to the actual velocity of Mars, as 1 is to 1.08. This excess of .08 of the force of gravity can be shown to have been expended, not in giving velocity to the earth, but on a very different object, thus leaving the actual velocity of the earth to its required velocity from gravity as 1 to 1. The same rule of procedure must be applied to all the other planets, except, of course, Neptune, which has no exterior planet; but in his case there is a special and peculiar way for showing how the comparatively large excess of gravity, thirty per cent., has been expended. All this appear as follows.

Hitherto all writers, except Sir William Herschel, who have tried to account for the rotation of a contracting nebulous mass, have in a vague, indefinite manner ascribed that rotation to the meeting of the particles at the centre of the mass. But we cannot look for the origin of the rotation in the central region of the mass, because that region is nearly without any motion. The entire central space, many millions of miles in diameter, is nearly quiescent, and all, or very nearly all, the movements must be on the surface of the mass. Let the line AB, divided into ten equal parts, represent a radius of a contracting globe, the centre being at A. Assuming that it contracts one-tenth part of its radius, every portion contracting equally, then 1 will move to 1', and 10 will move to 9. Thus it can be seen that the surface at 10 moves ten times more rapidly than the interior at 1. The nebular theory, as explained in this volume, regards the sun as having occupied all the space all around until half the distance to the nearest fixed stars; that is, on every side 3,622 times farther than the orbit of Neptune. But on the assumption that AB extended only 100 times farther than the orbit of Neptune, then the inward vertical movement at the orbit of the earth would be 3,000 times slower than at the surface of the nebulous mass. At the orbit of Mercury that movement would be 8,000 times slower than at the surface.

But there must be another and a far more efficient cause for rapid motion at the surface, and nearly at right angles to the former. The original nebular masses, breaking up from a nearly equal diffusion of matter through space, could not at first have been round. They must have been more or less irregular. Being expanded by the ordinary principle of repulsion which forms gases, the force of gravity could not cause a collapse or general unimpeded rush toward the centre. It could only bring down the prominences by lateral movements into the adjacent depressions. By this operation all around the globe very many currents would be formed, and these, by the composition of forces, would result in one general current.

This current could not be deep, except the original mass were extremely irregular. On account of the continued slow contraction of the nebulous globe, this current would move down an inclined plane, and therefore it could not stop, but must move more and more rapidly. This would constitute the rotation of the nebulous globe. But all the region inside of the rotating surface would move in radial lines toward the centre; and as we can conceive of nothing to turn those interior parts from the radial direction, so we can conceive of nothing to produce a rotation at the centre through mere contraction. The central region being filled up from all sides, would become the more dense, and that density would be an additional cause for a slower movement there. The rotation of the nebulous globe would, therefore, be wholly on the surface, and being always in frictional contact with the interior, it would be retarded by that interior. During the first stage of contraction in the history of a nebulous globe, the several partial currents would be retarded by one another, and perhaps, in some cases, be reduced to nearly *nil* before one general rotating current could be produced. Then, if that one current were very shallow, its retardation, by contact with the unrotating interior, would be the greater in proportion to its want of momentum, or *vis viva*. It is easy to conceive of cases where this initial current would be so shallow that its retardation would prevent it from ever acquiring a velocity that would make the centrifugal equal to the centripetal force, and hence such a sun could have no planets, or such a planet as Mars could have no satellite. Let us suppose, however, in the case of our sun, that when its surface, by contracting, had reached the orbit of Neptune, the velocity of its equatorial zone, impelled by gravity, was rapid enough to equalize the centripetal and centrifugal forces. Then a ring would be separated, and the velocity of the

APPENDIX II. 373

planet resulting from that ring would be equal to Neptune's. Therefore, in the case of the extreme outer planet, while coming down an inclined plane—say from infinite space— impelled all along by the force of gravity, we see how the entire amount of that force was not expended on the velocity of the planet; a large amount of the force was necessarily expended in overcoming resistances, when the several partial currents were counteracting one another, and when after all these had coalesced into one; this one primitive current was very shallow, and its small momentum was destroyed by frictional contact with the unrotating interior. By these resistances 30 per cent. of the velocity, which should have been produced by gravity, was destroyed. Hence we see how the actual velocity of Neptune, compared with the velocity of a fall from infinite space, is as 1 to 1.41.

Very different is the case of all the other planets and asteroids. From the orbit of Neptune the original nebulous sun contracted about 1,000,000,000 miles in radius to the orbit of Uranus. Then the surface of its equatorial zone, impelled by gravity down an inclined plane from one orbit to another, should have an additional velocity. This additional velocity should equal that of a fall from orbit to orbit. The velocity of this fall would amount to about one and a quarter miles per second; say 1.244. Add this to the actual velocity of Neptune, nearly three and a half miles per second—say 3.491—and we have 4.735 miles per second for the velocity of the equatorial zone at the orbit of Uranus. But the actual velocity of Uranus is 4.369 miles per second; and therefore the actual velocity of Uranus is to the calculated velocity from the force of gravity as 1 to 1.08. But on what has a portion of the force of gravity—the .08—been expended? Evidently it has been expended by friction on the unrotating interior of the nebulous sun. The discrepancy, therefore, can be accounted for. Moreover, if by friction there be a loss of velocity due to gravity of eight per cent. through a contraction in diameter of 2,000,000,000 miles, then there would be a loss of one per cent. through every 250,000,000 miles. This is an infinitesimally small loss, and nearly unworthy of being taken into account, even if we could not point out how it occurred. The same very near coincidence between the actual and the calculated velocities occurs in the case of every other planet and asteroid, as has been exhibited in two tables in their proper places.

Still, it may be said that "this relation proves nothing as to the origin of the orbital velocity." We do not see an actual connection between gravity as the cause and the orbital motion as the effect. The same may be said, and was long said, about gravity as the force that holds the planets and satellites in their orbits; and hence a long period elapsed, after Newton's demonstrations, before the scientific world in general believed gravity to be the centripetal force in astronomy. But the *mere coincidence* between the amount of gravity and the amount of centripetal force, occurring in so many cases, at last compelled belief. It is proper thus to advert to the grounds of our belief in gravity as the force that holds the planets in their orbits, in order to see whether we have the same grounds for belief in gravity as the force that gave them their velocities in their orbits. Here are two distinct classes of coincidences: the first proving the centripetal force, and the second proving the centrifugal or projectile force. The two classes are precisely equal in number and in significance. There is a third class of coincidences, adding force to the second, and proving that gravity is the origin of the rotations in the solar system as well as of the orbital revolutions. There is still a fourth class, also adding force to the second, and proving that gravity not only gave motion to the members of the solar system, but, while acting on a contracting nebulous mass, moulded and fashioned the solar system into an intricate piece of mechanism, consisting of more than a hundred separate parts, shaping and locating those parts in admirable harmony. Every satellite of Jupiter, as we have already said, has three simultaneous motions, one of rotation, one of revolution around Jupiter, and one of revolution around the sun; and not only these motions, but the separate existence of Jupiter apart from the sun, and the separate existence of each satellite apart from Jupiter, and the shapes of them all and their relative locations in space, coincide with the action of gravity on a contracting nebulous mass. Therefore, from this large number of coincidences, I think we have stronger grounds for belief in gravity, as the cause of the motions in the solar system, than in gravity as the centripetal force.

APPENDIX III.

THE VELOCITIES NECESSARY IN THE FIXED STARS TO PREVENT THEIR COMING IN COLLISION—TO XXXVIITH SECTION.

My methods for computing the motion necessary to prevent Alpha Centauri from falling into the sun were two, as follows:

FIRST METHOD.

I have already, in Appendix I., given the formula for ascertaining the velocity of a fall from infinite space: $V = \sqrt{2gr}$. Here g represents the velocity acquired at the end of one second, by a fall toward the sun, at the orbit of any planet. This is calculated from that at the surface of the earth, which is 32.16 feet. But if instead of the latter quantity we substitute the distance fallen through in one second, which is half as much, 16.08. then at any planetary orbit we obtain the velocity of a planet necessary to counteract the sun's centripetal force. Applying this to Alpha Centauri, we obtain the velocity in the text, or one hundred and forty-five miles per hour. Rapid as this may appear, it would yet take more than a million of years for one revolution in its imaginary orbit! The distance of Alpha Centauri was calculated from the parallax $0''.913$.

SECOND METHOD.

The other method is the well-known one of Kepler, namely:

As the cube of Neptune's distance
Is to the cube of Alpha Centauri's distance,
So is the square of Neptune's period
To the square of Alpha Centauri's period.

Neptune's distance was taken at 2,760,000,000 miles, and his period at 164.62 days. But these are well known to be only approximations, and hence the result was a motion of Alpha Centauri of 139.1 miles per hour; a very near coincidence, all things considered. A very slight alteration of Neptune's distance and period, within the limits of probable error, would have given the precise result of the first method.

APPENDIX IV.

CATALOGUE OF WRITERS ON THE CAUSE OF STELLAR MOTION, AND ON KINDRED SUBJECTS, QUOTED IN THE XLIST SECTION.

Sir Isaac Newton. Principia, 3d book; General Scholium. Letter to Burnett in Appendix 6 of his Life, by Sir David Brewster.

Sir William Herschel. Royal Philosophical Society's Transactions, year 1789; also, year 1811, pp. 284, 311, 319.

Pierre Simon, Marquis de Laplace. Système du Monde. Vol. 2.
George Louis Leclerc Buffon, quoted by Laplace in Système du Monde.
Emanuel Kant, quoted by Heinrichs, in American Journal of Science. Vol. 37.
Mary Somerville. The Connection of the Physical Sciences, Section 9.
Dionysius Lardner. Hand-Books of Natural Philosophy and Astronomy. Section 3,007.
J. P. Nichol. Professor of Astronomy in the University of Glasgow. His Encyclopædia of the Physical Sciences, Art. Nebular Hypotheses; and his Architecture of the Heavens, Letter 7.
Herman L. F. Helmholtz, Professor in the University of Heidelberg, Germany. London, Edinburgh, and Dublin Philosophical Magazine. Vol. 11, of 4th series, p. 503.
Stephen Alexander, Professor of Mathematics and Astronomy in Princeton College, New Jersey. Gould's Astronomical Journal. Vol. 2.
Vestiges of Creation; First Section, and Sequel to do., in the beginning.
Daniel Kirkwood, Professor of Astronomy in the University of Indiana. American Journal of Science. Vols. 30 and 38. Proceedings of the American Scientific Association, Second Meeting, p. 218.
Sears C. Walker, and
B. A. Gould, Jr. Proceedings of the American Scientific Association, Second Meeting, pp. 218 and 364.
François D. Arago, quoted by Alexander in Gould's Astronomical Journal. Vol. 2, p. 100.
Gustavus Heinrichs, Professor of Chemistry and Physics in the University of Iowa. American Journal of Science. Vols. 37–39.

David Trowbridge. American Journal of Science. Vols. 37–39.

O. M. Mitchell, Major-General in the United States Army, died at Hilton Head, in South Carolina, in the late war; formerly Director of the Astronomical Observatories at Cincinnati, Ohio, and Albany, N. Y. Popular Astronomy, and Astronomy of the Bible. Lecture 3d.
Pantecoulant, quoted fully by Mitchell.

APPENDIX V.

LAPLACE'S WRITINGS ON THE NEBULAR THEORY.

REGARDING the Nebular Theory as one of the most important productions of the human mind, and believing that many wrong ideas are afloat respecting its origin, I have thought proper to transcribe in this place all that its author ever wrote in bringing it before the world. The following extracts are from his "Système du Monde," as given in the translation of J. Pond, F. R. S. (London, 1809). The original was published in 1805; though of this I am not quite certain; and I have not at this moment the requisite authorities at hand. They are both from vol. ii.; the first from Book V., chap. vi., and the second from Book IV., chap. ix.

FIRST EXTRACT.

However arbitrary the system of the planets may be, there exists between them some very remarkable relations, which may throw light on their origin; considering them with attention, we are astonished to see all the planets move round the sun from west to east, and nearly in the same plane; all the satellites moving round their respective planets, in the same direction, and nearly in the same plane with the planets. Lastly, the sun, the planets, and those satellites in which a motion of rotation has been observed, turn on their own axes, in the same direction, and nearly in the same plane as their motion of projection.

A phenomenon so extraordinary is not the effect of chance; it indicates a universal cause which has determined all these motions. To approximate to the probable explanation of this cause, we should observe that the planetary system, such as we now consider it, is composed of seven planets and fourteen satellites. We have observed the rotation of the sun, of five planets, of the moon, of Saturn's ring, and of his farthest satellite; these rotations, with those of revolution, form together thirty direct movements in the same direction. If we conceive the plane of any direct motion whatever, coinciding at first with that of the ecliptic, afterwards inclining itself toward this last plane, and passing over all the degrees of inclination, from zero to half the circumference, it is clear that the motion will be direct in all its inferior inclinations to a hundred degrees,* and that it will be retrograde in its inclinations beyond that; so that, by the change of inclination alone, the direct and retrograde motions of the solar system can be represented. Beheld in this point of view, we can reckon twenty-nine motions, of which the planes are inclined to that of the earth, at most one-fourth of the circumference; but supposing their inclinations had been the effect of chance, they would have extended to half the circumference, and the probability that one of them would have exceeded the quarter would be $1 - \frac{1}{\text{\tiny?}}$, or $\frac{?????????}{?????????}$. It is then extremely probable that the direction of the planetary motion is not the effect of chance, and this becomes still more probable if we consider the inclination of the greatest number of these motions to the ecliptic is very small, and much less than a quarter of a circumference.

Another phenomenon of the solar system, equally remarkable, is the small eccentricity of the orbits of the planets and their satellites, while those of the comets are much extended. The orbits of the system offer no intermediate shades between a great and small eccentricity. We are here again compelled to acknowledge the effect of a regular cause; chance alone could not have given a form nearly circular to the orbits of all the planets; that cause must also have influenced the great eccentricity of the orbits of the comets, and what is very extraordinary, without having any influence on the direction of their motion; for, in observing the orbits of the retrograde comets as being inclined more than one hundred degrees to the ecliptic, we find that the main inclination of the orbits of all the observed comets approaches near to one

---

* Ninety degrees, by our mode of reckoning.

hundred degrees, which would be the case if the bodies had been projected at random.

Thus, to investigate the cause of the primitive motions of the planets, we have given the five following phenomena: 1st. The motions of the planets in the same direction, and nearly in the same plane. 2d. The motion of their satellites in the same direction, and nearly in the same plane with those of the planets. 3d. The motion of rotation of these different bodies, and of the sun in the same direction as their motion of projection, and in planes but little different. 4th. The small eccentricity of the orbits of the planets and of their satellites. 5th. The great eccentricity of the orbits of the comets, although their inclinations may have been left to chance.

Buffon is the only one whom I have known, who, since the discovery of the true system of the world, has endeavored to investigate the origin of the planets and of their satellites. He supposes that a comet, in falling from the sun, &c., &c. This hypothesis, then, is far from accounting for the preceding phenomena. Let us see if it be possible to arrive at their true cause.

Whatever be its nature, since it has produced or directed the motion of the planets and their satellites, it must have embraced all these bodies, and considering the prodigious distance which separates them, they can only be [it could only have been?] a fluid of immense extent. To have given in the same direction a motion nearly circular around the sun, this fluid must have surrounded that luminary like an atmosphere. This view, therefore, of planetary motion, leads us to think that in consequence of excessive heat the atmosphere of the sun originally extended beyond the orbits of all the planets, and that it has gradually contracted itself to its present limits, which may have taken place from causes similar to those which caused the famous star that suddenly appeared in 1572, in the constellation of Cassiopeia, to shine with the most brilliant splendor during many months.

The great eccentricity of the orbits of the comets leads to the same results; it evidently indicates the disappearance of a great number of orbits less eccentric, which indicates an atmosphere round the sun, extending beyond the perihelion of observable comets, and which, in destroying the motion of those which they have traversed in a duration of such extent, have reunited themselves to the sun. Thus we see that there can exist at present only such comets as were beyond this limit at that period. And as we can observe only those which in their perihelion approach near the sun, their orbits must be very eccentric; but,

APPENDIX V.

at the same time, it is evident that their inclinations must present the same inequalities as if the bodies had been sent off at random, since the solar atmosphere has no influence over their motions. Thus, the long period of the revolutions of comets, the great eccentricity of their orbits, and the variety of their inclinations, are very naturally explained by this atmosphere.

But how has it determined the motions of revolution and rotation of the planets? If these bodies had penetrated this fluid, its resistance would have caused them to fall into the sun. We may then conjecture that they have been formed in the successive bounds of this atmosphere by the condensation of zones, which it must have abandoned in the plane of its equator, and in becoming cold have condensed themselves toward the surface of this luminary, as we have seen in the preceding book. [See next extract.] One may likewise conjecture that the satellites have been formed in a similar way by the atmospheres of the planets. The five phenomena explained above naturally result from this hypothesis, to which the rings of Saturn add an additional degree of probability.

Whatever may have been the origin of this arrangement of the planetary system, which I offer with that distrust which every thing ought to inspire that is not the result of observation or calculation, it is certain that its elements are so arranged that it must possess the greatest stability, if foreign observations [influences?] do not disturb it.

SECOND EXTRACT.

All the atmospheric strata should take, after a time, the rotatory motion common to the body which they surround. For the friction of these strata against each other, and against the surface of the body, should accelerate the slowest motions, and retard the most rapid, till a perfect equality is established among them. In these changes, and generally in all those which the atmosphere undergoes, the sum of the products of the particles of the body and of its atmosphere, multiplied respectively by the area which their radii vectores projected on the plane of the equator describe round their common centre of gravity, are always equal in time.

Supposing, then, that by any cause whatsoever, the atmosphere should contract itself, or that a part should condense itself on the surface of the body, the rotatory motion of the body, and of its atmosphere, would be accelerated, because the radii vectores of the area, described by the particles of the primitive atmosphere becoming smaller, the sum of the product of all the particles by the corresponding area could not remain the same unless the velocity of rotation augments.

The point where the centrifugal force balances gravity, is so much nearer to the body in proportion as its rotatory motion is more rapid. Supposing that the atmosphere extends itself as far as this limit, and that afterwards it contracts and condenses itself from the effects of cold at the surface of the body, the rotatory motion would become more and more rapid, and the farthest limit of the atmosphere would approach continually to its centre; *it will then abandon successively in the plane of its equator fluid zones*, which will continue to circulate round the body because their centrifugal force is equal to gravity. But this equality not existing relative to those particles of the atmosphere distant from the equator, they will continue to adhere to it. It is probable that the rings of Saturn are similar zones abandoned by its atmosphere.

REMARKS.

The above reasoning about the comets is sound and admirable, but it was founded on wrong data, and has now no weight. The great error made by our author is contained in the italicized passage. It has never been detected either by the friends or by the opponents of the theory. The mere increase of angular velocity in consequence of contraction, with no increase in linear velocity, as contemplated by Laplace in his theory, would not add to the centrifugal force so as to cause a ring to be abandoned in the plane of the equator. Along with the increase of angular velocity, there must be an increase also of linear velocity. This we see in comparing any two of the planets, for instance the earth and Mercury; the one travelling 68,000 and the other 110,000 miles an hour. The great defect in Laplace's views, was in not perceiving that all the materials of the contracting and rotating atmosphere were moving down an inclined plane, and therefore subject to be hastened along by the force of gravity! Gravity acting on an inclined plane could alone increase the linear velocity so greatly as to make the centrifugal equal to the centripetal force, and thus cause the abandonment of rings in the plane of the equator. This was my first new idea on the nebular theory, received after many years of serious contemplation; and this, either directly or indirectly, has been my guide to all the others.

APPENDIX VI.

SINCE these sheets have been passing through the press, I have learned that M. Faye, in a Memoir recently communicated to the French Academy, has stated, as the results of his measurements, that the sun-spots are depressions beneath the sun's photosphere, varying in depth from about $\frac{1}{100}$th to $\frac{1}{100}$th of the sun's radius, *i. e.*, from about 40,000 to 20,000 miles. This differs widely from Wilson, and it is to be hoped that other investigators will enter on this matter.

NOTE TO SECTION XVII.

ON THE NECESSITY OF ROTATION.

1. THE necessity of rotation in a slowly contracting nebulous mass, and the force producing that rotation, are two ideas that must be kept clear and distinct, and at the foundation of all our investigations on the nebular theory; among other reasons, because the force producing the rotation is the force that gave motion to the stars. Mere contraction, the simple diminution of volume from loss of heat or from chemical action, could not produce rotation. In the general movement of contraction, "each atom would follow its leader toward the centre in radial lines," as has

been argued by the opponents of the nebular theory; and their argument has not been hitherto properly met.

2. Even with the force of gravity added to that of simple contraction, there would be no rotation if the velocity of contraction equalled the velocity due to the force of gravity. However irregular the shape of the nebulous mass, there would be only a direct central movement, and we see nothing to disturb or bend that movement.

3. And if the contraction were very slow, and the shape of the nebulous mass were perfectly round, then, with gravity added, there could be no rotation. We can conceive of nothing even in this case to hinder a direct central movement.

4. But if contraction were slow, slower than the motion due to gravity, and if the nebulous mass were uneven on its surface, then gravity would make the mass round and even, and in so doing would produce lateral or horizontal currents on the surface; *these surface currents, impelled by gravity, would run down an inclined plane as long as the mass contracted;* therefore they would continue, and, by the composition of forces, they would unite in one general current, and this one current would form the rotation of the nebulous globe. Although but a surface current at first, still, by friction, it would at last penetrate to the centre.

5. Thus it appears that gravity alone is the force that produces rotation in a slowly contracting nebulous mass, and therefore this is the force that alone gave motion to the stars.

NOTE TO SECTION XXIX.

THROUGH inadvertency, and from not having Mayer's treatise at hand, I have given an inaccurate and too favorable a view of the operation of his theory for the retardation of the moon's rotation. Instead of a single tide-wave, which retarded the moon's rotation, he rightly supposes another tide-wave on the opposite side of that orb, which tended to accelerate the rotation, but in a smaller degree, owing to its greater distance from the earth. Hence the net amount of retardation was the difference of the earth's gravitating force on these two waves, which must have been small, but still effective. In my own view, retardation was caused by rigidity operating against both waves. I am anxious not to offer even the appearance of detracting from the great merit of Mayer. We can never too much admire his patient, persevering, and laborious thinking, when he discovered, from abstract views, the mechanical equivalent of heat, and laid the foundation for the grand theory of the conservation of force.

PERSONAL INDEX.

A
Abbott, 135.
Agelander, 105, 106, 124.
Airy, 84, 90, 144.
Alexander, 336, 377.
Arago, 20, 131, 338, 377.

B
Bayer, 105.
Berard, 133, 140.
Bessel, 122.
Biot, 286.
Bode, 257.
Brewster, 376.
Buffon, 334, 377, 380.
Burnett, 376.

C
Carrington, 97.
Clark, 24, 141, 313.
Copernicus, 213, 313, 344, 346.

D
Darwin, 356, 359.
Davy, 4, 79, 80, 347.
Dawes, 86.
Dembowski, 135, 146–150.
Dick, 90, 91.
Donati, 134, 135, 137, 138, 140.
Draper, 191.

E
Espy, 294.
Euler, 286.

F
Fahrenheit, 294.
Faraday, 179, 192, 225.

Faye, 383.
Fergani, 133, 140.
Fletcher, 146.
Fraunhoffer, 18, 162, 175.

G
Gilliss, 133, 140.
Goldschmidt, 24.
Gould, 337, 338, 377.

H
Heinrichs, 333, 334, 338, 343, 377.
Hels, 133.
Helmholtz, 335, 377.
Henry, 90.
Herschel, W., 23, 24, 121, 146–149, 162, 332–334, 338, 340, 341, 370, 376.
Herschel, J. F. W., 88, 89, 91, 92, 108, 122, 124, 134, 135, 140, 146, 149, 199, 302, 304, 334.
Hipparchus, 123.
Humboldt, 105, 132–135, 137, 138, 140, 141, 142, 159, 160, 199.

J
Jacob, 148.

K
Kant, 334, 377.
Kearney, 133, 134, 138, 140.
Kepler, 237, 338, 344, 376.
Kirchoff, 18, 175.
Kirkwood, 337, 343, 377.

L
Lagrange, 286.

PERSONAL INDEX.

Laplace, 4, 79, 122, 289, 292, 317, 333, 334, 338, 340, 341, 343, 345–347, 377, 382.
Lardner, 335, 343, 377.
Lassell, 26.
Laugier, 89, 92.
Lexell, 294.
Long, 92.
Lyell, 352, 353, 359, 363.

M
Mackay, 133, 140.
Marriotte, 136.
Mayer, 77, 78, 179, 269–271, 273, 384, 385.
Milton, 3, 217.
Mitchell, Maria, 129, 135, 145.
Mitchell, 339, 340, 343, 378.
Murray, 148.

N
Newton, 4, 79, 179, 213, 288, 291, 317, 326–329, 331, 338, 343, 344, 347, 359, 374, 376.
Nichol, 335, 336, 341, 343, 377.

P
Pantecoulant, 340, 378.
Pond, 378.
Pouillet, 294.
Ptolemy, 133, 140.

R
Riccioli, 133, 140.
Ross, 94.
Rumkoff, 175.
Rutherfurd, 176.

S
Schmidt, 134, 138, 140, 143, 147.
Schonbein, 192.
Schwabe, 89, 93.
Secchi, 148–150, 175.
Seneca, 134, 140.
Sestini, 135, 144, 145, 146, 149, 150.
Smyth, 145, 147, 149.
Smyth, C. Piazzi, 144, 148.
Sommerville, 334, 377.
South, 146, 147.
Struve, 121, 144–150, 162.

T
Trowbridge, 338, 341, 378.
Tyndall, 362.

W
Walker, 337, 377.
Webb, 135, 145–150.
Wilcocks, 134, 140.
Wilson, 82, 383.
Wollaston, 92.

INDEX OF STARS.

Aldebaran, 27, 24, 129, 133, 136, 139, 152, 156.
Algol, 105, 108, 112.
Altair, 137, 139, 142, 156.
Andromedæ, Alpha, 26.
" Gamma, 148, 296.
Antares, 21, 24, 25, 129, 133, 139, 152.
Aquarii, R., 105.
Aquilæ, Eta, 105, 109, 113.
Arcturus, 24, 133, 139, 143, 152, 156.
Argus, Eta, 121, 124, 125, 133, 139.
Arietis, Gamma, 148.
Astræa, 230.
Aurigæ, Epsilon, 105, 109, 115, 120.
Bellatrix, 136, 156.
Betelguese, or Alpha Orionis, 105, 108, 110, 129, 133, 136, 139, 152, 156.
Bootis, Iota, 146.
" Mu, 26.
" Pi, 146.
" Xi, 145.
39 Bootis, 146.
Cancri, R., 105.
" S., 105.
" Iota, 146.
" Xi, 26.
" Zeta, 20, 24, 25.
Canopus, 121, 140.
12 Canum Venaticorum, 149.
Capella, 133–136, 136, 137, 143, 156, 160.
Cassiopeia, 76, 123, 156, 380.
" Alpha, 105, 114.

P. II. Cassiopeiæ, 148.
Castor, 135, 136, 296.
Centauri, Alpha, 140, 285, 297, 312, 375, 376.
Centauri, Beta, 140.
Cepheus, 123.
Cephei, Delta, 105, 109, 110, 113.
" Kappa, 148.
" Zeta, 149.
Ceti, Gamma, 148.
" Mira, 105, 107, 109, 110, 114, 115, 166, 211.
Comæ Berenices, 305.
Coronæ R., 105, 108, 114, 121, 124.
" Sigma, 149.
Corvi, Delta, 146.
Crucis, Alpha, 133, 139.
" Kappa, 135, 160, 305.
Cygnus, 108, 124.
Cygni, Alpha, 137, 142.
" Chi, 105, 115.
" Gamma, 147.
" Mu, 150.
61 Cygni, 22.
Deneb, 137, 142, 156.
Denebola, 157.
40 Draconis, 149.
Earth, 199, 200, 203, 224, 230, 234, 236, 245, 248, 252, 254, 258, 261, 262, 265, 277, 288, 322.
Eridani, Alpha, 140.
32 Eridani, 145.
Fomalhaut, 157.
Geminorum, Upsilon, 211.
" Zeta, 105, 110, 113.
Herculis, Alpha, 105, 107, 114, 146.
95 Herculis, 143.

INDEX OF STARS.

Hydra Hevelli, 105, 115, 120.
Hydræ, Alpha, 105.
Hygeia, 230.
Jupiter, 11, 16, 21, 27, 122-124, 160, 173, 184, 199, 200, 224, 230, 234, 237, 244, 245, 248, 249, 252, 254-256, 258, 261, 262, 264-269, 276, 277, 282, 283, 288, 292, 293, 295, 296, 298, 304, 307, 322, 325, 373.
Leonis, Gamma, 148.
" R., 105, 114.
Lupi, Mu, 26.
12 Lyncis, 26.
Lyræ, Alpha, 135.
" Beta, 105, 109, 113, 118, 147.
Lyræ, Eta, 26, 148.
" Epsilon, 147.
" Gamma, 147.
Magellanic Clouds, 309.
Mars, 18, 27, 132-134, 136, 142, 199, 230, 234, 236, 245, 250, 252, 254, 259, 261, 262, 277, 287, 370, 372.
Mercury, 16, 21, 27, 173, 184, 199, 202, 229-231, 233-236, 245, 250, 252-254, 257, 259, 261, 262, 264, 267, 277, 282, 287, 269, 290, 299, 322, 329, 330, 346, 360, 371, 382.
Mira Ceti, 105, 107, 109, 110, 114, 115, 166, 211.
Mizar, 145.
11 Monocerotis, 26.
Neptune, 16, 199, 202, 227, 229-231, 234-238, 241, 245, 248, 249, 251, 252, 254, 258, 260-262, 265, 268, 274, 277, 299, 329, 330, 346, 348, 360, 370-373, 375, 376.
Ophiuchus, 124.
39 Ophiuchi, 145.
Orionis, Alpha, 105. *See* Betelguese.
23 Orionis, 145.
Orionis, Theta, 26, 174.
" Trapezium, 174, 297.
Pallas, 297, 301.
Pegasi R., 105.

Pegasi, Beta, 105.
Piscium, Alpha, 150.
25 Piscium, 145.
Pleiades, 305.
Polaris, 144, 156, 157.
Pollux, 137, 139.
Præsepe, 305.
Procyon, 21, 24, 25, 134, 136, 139, 156.
Regulus, 157.
Rigel, 21, 24, 134, 136, 139, 156.
Saturn, 16, 27, 133, 173, 184, 199, 200, 202, 224, 230, 234, 245, 248, 249, 252, 254-258, 260-262, 264-269, 276, 277, 282, 283, 295, 296, 304, 307, 321, 322, 325, 329, 379.
Sagittarius, 123.
Scorpio, 123.
Scorpii, Sigma, 146.
Scuti R., 105, 107, 114.
Serpentis R., 105.
" S., 105.
" Delta, 146.
Sirius, 21, 24, 25, 121, 123, 125, 132, 134, 136-139, 141-143, 152, 156, 285, 313.
Southern Cross, 309.
Spica, 136, 137, 139, 156.
Trianguli, Iota, 148.
Uranus, 184, 199, 200, 202, 230, 231, 234-236, 241, 245, 248, 251, 252, 254, 257, 258, 260-262, 265-267, 261, 282, 304, 322, 324, 325.
Ursa Major, 156, 157.
Ursæ Majoris, Zeta, 145.
Ursæ Minoris, Alpha, 146.
" " Beta, 133.
Vega, 21, 135, 136, 138, 139, 141, 142, 156, 160.
Venus, 11, 16, 122-124, 160, 199, 200, 202, 230, 231, 234-236, 245, 250, 252, 254, 257, 259, 260, 262-264, 277, 282, 298, 299, 322.
Virginis R., 105, 114.
" Gamma, 145.
Vulpes, 124.

GENERAL INDEX.

A

Artesian wells, 30.
Astral systems. *See* Sidereal Systems.
Asteroids, their origin, 292; eccentricities and inclinations of orbits, 287.
Atols, 37, 38.
Attraction, atomic, 13, 22.

C

Central sun, 306, 307; nebulous stars are central suns, 301; a central sun not necessary, 309; no central sun in our own sidereal system, 309.
Centrifugal force in astronomy caused by gravity, 225.
Chemical action causes the light and heat of the sun, 170; conditions of chemical action in the sun, 176; facts explained by, 206-213.
CHEMICAL ELEMENTS, 57; their tendency to combine, 57; their combination in our globe was rapid and violent, 59; they were not originally created in combination, 60, 65; in the sun, 81; very different in the sun and stars, 175; they are mere modifications of the same fundamental matter, 182, 187, 189, 190; they were formed during the nebulous condensations, 193, 199, 200, 203; they are now forming in the sun, 202.

CHEMICAL FORCE unites chemical elements, 63, 65, 69; its laws, 64; its extension through all space, 103; the origin of chemical force, 214, 218, 348-351.
Clusters of stars, 279.
Coincidences between theory and fact, 374.
Colored stars, 128; changes in their colors, 131; causes of change in colors, 130, 131, 137, 158, 162; catalogue of changes, 132; rules for observing colored stars, 154; influence of earth's atmosphere on colored stars, 155.
Comets, 294.
Creation, modern theories of, 3, 343; a slow process, and always from simple to complex, 60; Copernican, 344; gravity necessary for the Copernican theory, 344; nebular theory, 345; gravity necessary for the nebular theory, 346; creation of light, 4, 347; the force in creation necessary to prolong the light of the sun, 348; the geological theory of creation, 352; the forces in the geological theory, 352; the theory of the creation of plants and animals, or transmutation theory, 353; the force in the creation of species, 356; theories of operation and theories of force, 358; history of creation, 359; present prog-

ress of creation, 364, 365; creation effected by special forces, 365; creation the work of Design, 366; these forces in creation have been designed and appointed by Divine Providence, 366.

D

Days, length of, why unaltered, 33, 273.
Density, law of, in solar system, 202.
Disk of nebulous matter, 299; its separation into stars, 300, 302; our immediate sidereal system is a disk, 302, 303, 305.
Distances of planets from the sun, 13, 14, 256–259; why distances between planets greater than between satellites, 267.
Double and multiple stars, 295; they were formed in accordance with the nebular theory, 296.

E

Earth, once self-luminous, 28, 29; its density, 32; rising and falling of its crust, 37, 47–49; its oblateness, 38; greatness of its features, 43; former high temperature of surface, 44; chemical elements of, 185; motion of its axis of rotation, 50.
Earthquakes, 32; causes of, 44.
Eccentricities of orbits of planets, 287; of asteroids, 287.

F

Faculæ, or ridges of flames in the sun, 86, 87.
Falls, from infinite space, velocity of, 228, 230.
Faults, 37.
Fixed stars. *See* Stars.

G

Gravity, the cause of stellar motion, 317–320; facts accounted for by gravity, 321; gravity compared in Copernican and nebular theories, 326.

H

History of creation, 359. *See* Creation.

I

Inclined plane, 226, 227.
Inclination of axes of rotation, and of equatorial and orbital planes, 278; of orbits of asteroids, 287.
Irregular stars, 120.

L

Land, how formed, 39; geological history of, 46; rising and falling of, 37, 47–49; why more in northern than southern hemisphere, 49.
Lateral pressure, 38.
Life, history of vegetable and animal, 51, 363.
Lost stars, 122.

M

Man, his antiquity on our globe, 363.
Matter in a nebulous condition, how many times more rare than hydrogen, 348; a connecting link between ponderable and imponderable matter, 245; not diffused by heat, 194, 219; its contraction caused by chemical action, 193–195.
Meteorites, 186, 201, 292.
Milky Way, 302; it is a ring, 308; we are near its centre, 309.
Mines, 29.
Moon, why so slow in movement, why so large, why so far from the earth, 269; why only one rotation in one revolution, 270, 271; its igneous history, 71, 165; its chemical elements different from those on the earth, 184.
Motion, proper, of fixed stars, 310, 311; cause of, 316; opinions of scientific men of, 331.
Mountains, how formed, 39; why parallel to each other and to coast lines, 40, 41; why through the centres of long islands and

GENERAL INDEX.

peninsulas, 41; why their positions on high table-lands, 43; why the abodes of thermal springs and volcanoes, 40.
Mundane system, peculiarities of, 268.

N

Nebulæ, various forms of, 299; planetary, 301; elliptical, 301; annular, 302; irregular, near the Milky Way, 303, 304.
Nebular theory, 193-195, 216, 217; its origin by Laplace, 345; its former deficiency was the want of a force to impel its operations, 346; Laplace's writings on, 378; remarks on, 382.
Nebulous matter. *See* Matter.
Nebulous stars, 301.

O

Oblateness, the origin of, 224; influence of the oblateness of Jupiter and Saturn, 282; of the sun, 283, 287-289; loss of sun's oblateness changed the planetary orbits from circles to ellipses, 289, 290.

P

Periodic stars, 104; table of, 105; causes of their periodicity, 106.
Planets, their resemblance to our earth, 12; their diameters, 12; they have the forces of gravitation, atomic attraction and repulsion, and of inertia, 13; their distances from the sun, 13, 356; relation between their periods and distances, 14; their motions, 14; their orbits, 14; major and minor axes of orbits, 15; their densities, 16; they are of the same nature as the sun and fixed stars, 28; they have very different chemical elements, 184.
Pores, 89, 90.

R

Repulsion, atomic, 13, 22, 195, 217.

17*

Retardation, 230-238, 249, 308; it has lengthened our days, 254.
Retrograde, rotation of Uranus, 281.
Rings, nebulous, separated by centrifugal force, 239-241; dimensions of rings, 240, 276, 277; their breaking, 241; the rings parted only from a narrow equatorial belt, and consequently very thin, 240.
Rocks, igneous, 34; aqueous, 35.
Rotation, the necessity of, 220; velocity of, 226, 227; at first rotation was only on the surface, 230-233; why from west to east, 274; why only one rotation of the moon in one revolution, 270, 271; why the sun's rotation more retarded than that of the planets, 255.
Routines of irregular variations in the light of fixed stars, 108, 109, 110-112.

S

Satellites, of the same nature as planets, 26, 27, 243; their nebulous sizes, 244, 245; why belong only to the larger planets, 261; table of their actual velocities compared with their theoretical velocities, 248.
Saturnian system, 264.
Sidereal systems, various forms, 299; our own, 302-305; cause of the diversities among sidereal systems, 298, 306.
Solar system, its formation, 241-243, 318-320; facts in the solar system, whose origin is accounted for by gravity, 321.
Spots in the sun, 76, 90; their nuclei and penumbra, 90; they are near ridges of flames, 90; deficient in heat, 90; numbers of, 90, 96; changes in, 91; cracks or fissures in, 92; their breaking in pieces and flying asunder, 92, 98; the lines defining the spots, 92, 93; bright

border around the spots, 93;
nature of spots, 93, 94; effect
of the spots when large, 94;
why they disappear, 95; they
cause vertical currents, 96; cycles of about eleven years in
their numbers, 96; they seem
dark only by contrast, 90, 97.
Springs, thermal or hot, 31; why
more often in mountains, 40.
Stars, fixed, they are suns, 17;
they rotate, 18; their prismatic
spectra, 18; surrounded by
flames, 20; have liquid bodies,
20; have planets, 20, 24; their
proper motions, 22; their mutual influences by gravity, 294,
255, 313, 314; have simple
chemical elements, 102; various
classes of fixed stars, 104; periodic stars, 104; cause of their
periodicity, 106, 115, 116; their
maxima and minima of light,
107; irregularities of increase
and decrease in light, 108; irregularities of their periods, 111;
routines of irregularity, 108;
changes of their routines, 108,
109; connection between changes
of routine and changes of period,
112; why the increase of their
light is more rapid than the decrease, 116, 117; chemical elements different in different stars,
183; systems of stars, diversities
of, 298, 305–308. See Colored
Stars; also Temporary, Lost,
Periodic, and Variable Stars;
Double and Nebulous Stars.

Sun, theories of, 73; envelope theory, 72; meteoric theory of sun's
heat, 77; chemical theory of, 78,
79, 80, 109; constitution of sun,
80, 160; the sun a variable star,
106, 108, 115; surrounded by
an envelope of flames, 81; these
flames in violent agitation, 82;
the sun an incandescent liquid,
83; sun's atmosphere, 78, 85;
clouds in sun's atmosphere, 84;
their height and breadth, 85;
why not ordinarily visible, 85;
heat of, 87; light of, 88. See
Spots, Pores, Faculæ. Currents in, 97; proof of local
currents, 97; want of density,
202; chemical elements now
found in, 202.
System. See Mundane, Saturnian,
Solar, Sidereal.

T

Temporary stars, 123; cause of,
125.
Trap rocks, nature of, 35; dykes
of, 36.

V

Variable stars, irregular, 120, 121.
Velocity, inconceivably slow, only
one degree in twelve million
years, 290.
Volcanoes, 31; why more often
in mountains, 40; cause of, 44;
pseudo-volcanoes in the United
States beyond the Mississippi,
12

THE END.

www.ingramcontent.com/pod-product-compliance
Lightning Source LLC
Chambersburg PA
CBHW032015220426

43664CB00006B/256